AFFINITY CHROMATOGRAPHY

CHROMATOGRAPHIC SCIENCE

A Series of Monographs

Editor: JACK CAZES
Silver Spring, Maryland

Volume 1: Dynamics of Chromatography (out of print)
J. Calvin Giddings

Volume 2: Gas Chromatographic Analysis of Drugs and Pesticides
Benjamin J. Gudzinowicz

Volume 3: Principles of Adsorption Chromatography: The Separation of Nonionic Organic Compounds (out of print)
Lloyd R. Snyder

Volume 4: Multicomponent Chromatography: Theory of Interference (out of print)
Friedrich Helfferich and Gerhard Klein

Volume 5: Quantitative Analysis by Gas Chromatography
Joseph Novák

Volume 6: High-Speed Liquid Chromatography
Peter M. Rajcsanyi and Elisabeth Rajcsanyi

Volume 7: Fundamentals of Integrated GC-MS (in three parts)
Benjamin J. Gudzinowicz, Michael J. Gudzinowicz, and Horace F. Martin

Volume 8: Liquid Chromatography of Polymers and Related Materials
Jack Cazes

Volume 9: GLC and HPLC Determination of Therapeutic Agents (in three parts)
Part 1 edited by Kiyoshi Tsuji and Walter Morozowich
Part 2 and 3 edited by Kiyoshi Tsuji

Volume 10: Biological/Biomedical Applications of Liquid Chromatography
Edited by Gerald L. Hawk

Volume 11: Chromatography in Petroleum Analysis
Edited by Klaus H. Altgelt and T. H. Gouw

Volume 12: Biological/Biomedical Applications of Liquid Chromatography II
Edited by Gerald L. Hawk

Volume 13: Liquid Chromatography of Polymers and Related Materials II
Edited by Jack Cazes and Xavier Delamare

Volume 14: Introduction to Analytical Gas Chromatography: History, Principles, and Practice
John A. Perry

Volume 15: Applications of Glass Capillary Gas Chromatography
Edited by Walter G. Jennings

Volume 16: Steroid Analysis by HPLC: Recent Applications
Edited by Marie P. Kautsky

Volume 17: Thin-Layer Chromatography: Techniques and Applications
Bernard Fried and Joseph Sherma

Volume 18: Biological/Biomedical Applications of Liquid Chromatography III
Edited by Gerald L. Hawk

Volume 19: Liquid Chromatography of Polymers and Related Materials III
Edited by Jack Cazes

Volume 20: Biological/Biomedical Applications of Liquid Chromatography IV
Edited by Gerald L. Hawk

Volume 21: Chromatographic Separation and Extraction with Foamed Plastics and Rubbers
G. J. Moody and J. D. R. Thomas

Volume 22: Analytical Pyrolysis: A Comprehensive Guide
William J. Irwin

Volume 23: Liquid Chromatography Detectors
Edited by Thomas M. Vickrey

Volume 24: High-Performance Liquid Chromatography in Forensic Chemistry
Edited by Ira S. Lurie and John D. Wittwer, Jr.

Volume 25: Steric Exclusion Liquid Chromatography of Polymers
Edited by Josef Janča

Volume 26: HPLC Analysis of Biological Compounds: A Laboratory Guide
William S. Hancock and James T. Sparrow

Volume 27: Affinity Chromatography: Template Chromatography of Nucleic Acids and Proteins
Herbert Schott

Volume 28: HPLC in Nucleic Acid Research: Methods and Applications
Edited by Phyllis R. Brown

Volume 29: Pyrolysis and GC in Polymer Analysis
Edited by S. A. Liebman and E. J. Levy

Volume 30: Modern Chromatographic Analysis of the Vitamins
Edited by Andre P. De Leenheer, Willy E. Lambert, and Marcel G. M. De Ruyter

Volume 31: Ion-Pair Chromatography
Edited by Milton T. W. Hearn

Volume 32: Therapeutic Drug Monitoring and Toxicology by Liquid Chromatography
Edited by Steven H. Y. Wong

Volume 33: Affinity Chromotography: Practical and Theoretical Aspects
Peter Mohr and Klaus Pommerening

Other Volumes in Preparation

AFFINITY CHROMATOGRAPHY

Practical and Theoretical Aspects

PETER MOHR and KLAUS POMMERENING

Central Institute of Molecular Biology
Academy of Sciences of GDR
Berlin-Buch, German Democratic Republic

WITH CONTRIBUTIONS BY

BERND EBERT, THOMAS HANKE,
WOLF-HAGEN SCHUNCK
Central Institute of Molecular Biology
Academy of Sciences of GDR

AND WERNER SCHÖSSLER
Central Institute of Cardiovascular Research
Academy of Sciences of GDR

Marcel Dekker, Inc. **New York and Basel**

Library of Congress Cataloging in Publication Data

Mohr, Peter, [date]
Affinity chromatography.

(Chromatographic science ; v. 33)
Includes bibliographies and index.
1. Affinity chromatography. 2. Biomolecules--
Analysis. I. Pommerening, Klaus, [date]
II. Title. III. Series.
QP519.9A35M64 1986 574.19'285 85-25262
ISBN 0-8247-7468-X

MARCEL DEKKER, INC.
270 Madison Avenue, New York, New York 10016

Current printing (last digit):
10 9 8 7 6 5 4 3 2 1

PRINTED IN THE UNITED STATES OF AMERICA

Preface

The general principles of affinity chromatography as a method carried out in liquid phase have been developed over the last two decades. An essential premise for the elaboration of this method was the increasing knowledge of the structure of biochemical systems and of the fundamentals of their interactions in living matter. Investigations in this field stimulated us and many others to concentrate on selected problems in this field.

In the meantime, numerous biologically active substances previously obtained with much difficulty could be isolated in a simple manner by means of affinity chromatography. Further, substances were for the first time proved or prepared by this method. Although affinity chromatography currently is used as a preparative method mainly on the laboratory scale, it can be assumed that its principles will be applied more and more for scale up preparations and for the routine analysis of biologically active substances. These successful developments and the high potentiality for growth in the field of affinity chromatography have resulted in a continually increasing body of relevant literature. It is the aim of the authors, therefore, to summarize in this short monograph present knowledge of this field as completely as possible and to give a survey of its current state of development, including such varieties of affinity chromatography as metal chelate, charge transfer adsorption, hydrophobic interaction, and covalent chromatography. Furthermore, some special methods are discussed as well. These include dye-ligand chromatography, the affinity chromatographic separation of membrane ingredients, and affinity electrophoresis.

Numerous colleagues have lent helpful support to our project. We would like to express our great appreciation to Dr. Bernd Ebert and Mr. Thomas Hanke for preparation of part of Chapter 6, Dr. Wolf-Hagen Schunck for preparation of Chapter 11, and Dr. Werner Schössler for contributing Chapter 18. We hope that this cooperation has enabled us to present an extensive and profound treatment of the topics contributed by these colleagues. Finally, we are indebted to Prof. Dr. F. Jung for critical reading of the manuscript, Mr. H. Marquardt for linguistic corrections, and Ms. G. Jonas for the drawings. Moreover, we wish to express our gratitude to the staff at Marcel Dekker for all their efforts.

We hope that this survey on affinity chromatography will be a helpful contribution towards clearing up some of the main problems faced by those involved in the field of liquid chromatography, and also that it will provide stimulation for prospective research work.

Peter Mohr
Klaus Pommerening

Contents

AFFINITY CHROMATOGRAPHY

part I
INTRODUCTION

1
Introductory Remarks

Our increasing insight into the structure and function of living matter yields evermore practical results. Impressive examples emphasizing this statement are provided by biospecific affinity chromatography, other affinity chromatographic methods, and their related techniques. All these methods are varieties of liquid chromatography (Table 1.1) that have gained an essential importance for the separation, isolation, purification, and analytic characterization of various kinds of high- and low-molecular weight compounds both in the laboratory and on an industrial scale. By reason of favorable experimental working conditions (e.g., normal temperature and aqueous solution), these methods are also qualified to a high degree for separation, concentration, and characterization of numerous more or less sensitive biologically active substances. The enormous progress in biochemistry in the last decades, therefore, is closely connected with the development of such liquid chromatographic procedures.

After introducing adsorption, ionic exchange (Peterson and Sober, 1956), and gel permeation chromatography (Porath and Flodin, 1959) for use with biological material, the various types of affinity chromatography have been developed continuously in the last few years and applied in laboratory practice. Compared with other liquid chromatographic methods, these techniques are eminently qualified for high specific separation, isolation, and purification of native biologically active substances, as well as their artificial equivalents, from more or less large amounts of contamination. Furthermore, an essential advantage is their concentrating effect, which allows the isolation of a desired substance also from larger sample volumes. The basis of the efficiency of affinity chromatographic procedures is a reversible,

TABLE 1.1 Methods of Liquid Chromatography

Chromatographic method	Basis of the separation effect	Examples of separable substances
Distribution	Different solubility of a substance in two unmixable liquid phases	Amino and carbonic acids, lanthanides
Adsorption	Unspecific adsorption on solid-phase supports, such as aluminum oxide or silica gel	Compounds and substances of varied structure
Ion-exchange	Adsorption by electrostatic interactions	Low- and high-molecular-weight charged compounds, such as peptides, proteins, polynucleotides, hormones
Gel permeation	Molecular sieve effects of the adsorbent particles	Natural, semisynthetic, and synthetic compounds of different molecular weights (e.g., peptides, proteins, polysaccharides, polynucleotides, hormones)
Affinity[a]	Biospecific sorption, group-specific interactions, chemisorption	See Chap. 2, Table 2.1

[a]Frequently this method is not included in the liquid chromatographic procedures since separation of a medley of substances in one step is not attainable in many cases.

more or less specific, interaction between a solid phase bound group (generally termed "ligand") and the substance to be separated (termed "affine component").

REFERENCES

Peterson, E. A., and Sober, H. A. (1956). Chromatography of proteins. I. Cellulose ion-exchange adsorbents. *J. Amer. Chem. Soc. 78*; 751-755.

Porath, J., and Flodin, P. (1959). Gel filtration: A method for desalting and group preparation. *Nature 183*: 1657-1659.

2
History of Affinity Chromatography

The fundamental principle of binding a biologically active substance using a specific and reversible interaction to a suitable matrix traces back to Starkenstein (1910), who used insoluble starch and the enzyme α-amylase. Starting with investigations on antigen-antibody interactions using invertase antibodies adsorbed to kaolin, Engelhardt had proposed in 1924 the "principle of the fixed partner" as a general method for the isolation of biologically active substances (Engelhardt et al., 1970). It was not until 1951 that Campbell and colleagues described the purification of antibodies by antigens chemically attached to cellulose. Rapid development of affinity chromatography and related techniques began after the introduction of the Sephadex gels by Porath and Flodin (1959), as well as the beaded agaroses by Hjerten (1962), as suitable matrix materials. With the finding that molecules containing primary amino groups can be coupled on polysaccharide matrices activated by CNBr (Axen et al., 1967), an essential prerequisite was created for the development of numerous affinity chromatographic procedures and their introduction into laboratory practice. This process was further promoted as Cuatrecasas and coworkers (1968) succeeded in solving the problem of steric accessibility by interposition of a chain termed "spacer arm" between the ligand and the solid-phase support.

Enormous broadening of the applicability of affinity chromatographic principles was afforded by the development of further support materials, including Spheron (Coupek et al., 1973), Enzacryl (Epton et al.,

1974), beaded cellulose, and polymer-coated inorganic materials, as well as new techniques for the introduction of ligands and spacer arms in matrix materials (Chaps. 3 and 4).

Besides the preparation and characterization of biologically active compounds, affinity chromatography is also receiving more and more interest as a potential method for the separation of viruses (Kristiansen, 1975) and cells (Hayman et al., 1973; Sharon, 1979).

Several authors have defined affinity chromatography as a method based on specific and reversible molecular interactions between biologically active substances (Cuatrecasas and Wilchek, 1968). In the following chapters the term "affinity chromatography" will be used more generally as proposed by Porath (1982). This means that several variants are involved, even when the matrix-bound ligand does not originate exclusively from biological matter. Examples are metal chelate chromatography (Porath, 1978), charge transfer adsorption chromatography (Porath and Dahlgren-Caldwell, 1977), hydrophobic interaction chromatography (Yon, 1972), dye-ligand chromatography (Easterday and Easterday, 1974), and, in a qualified sense, covalent chromatography (Table 2.1) (Brocklehurst et al., 1973). In all these cases affinity chromatographic efficiency is due to chosen groups anchored as ligands on the matrix material. A simplified representation illustrating the different modes of interaction between ligands and affine components on a molecular level is given in Figure 2.1. Furthermore, there is no reason to avoid using the term "affinity chromatography" when the substance to be separated by means of one of these techniques is not of biological origin.

The continuous deepening of our understanding of steps and processes connected with affinity chromatographic procedures has increasingly provided starting points for the development of numerous modified techniques resting on group-specific interactions and biological recognition principles. Examples are the affinity partition technique (Flanagan and Barondes, 1975), high-performance liquid affinity chromatography (Ohlson et al., 1978), affinity electrophoresis (Bøg-Hansen, 1973), the analytic application of immunotechniques (Yalow and Berson, 1959; Aizawa et al., 1980), affinity chromatography at subzero temperatures (Balny and Douzou, 1979), the affinity elution technique (Scopes, 1977a, 1977b, 1982), the affinity electrode principle (Lowe, 1979), affinity therapy (Ringsdorf, 1975), and affinity precipitation (Larson and Mosbach, 1977), as well as magnetic affinity chromatography (Table 2.2) (Mosbach and Andersson, 1977).

TABLE 2.1 Variants of Affinity Chromatography

Chromatographic method	Basis of the separation effect	Examples	References
Biospecific affinity (biosorption)	Specific adsorption (biospecific recognition)	Biologically active substances, including cell fragments, viruses, and cells	Cuatrecasas et al. (1968)
Dye-ligand	Specific interaction of biomacromolecules with triazine and triphenylmethane dyes	Enzymes (dehydrogenases, kinases, esterases, peptidases), nucleic acids, nucleic acid binding proteins	Easterday and Easterday (1974)
Metal chelate	Complex formation between a matrix-bound metal chelate and the substance to be separated by exchange of low-molecular-weight metal-bound ligands	Peptides, proteins, nucleic acids	Porath (1978)
Charge transfer adsorption	Interaction between electron-accepting and electron-donating groups	Amino acids, peptides, nucleotides	Porath and Dahlgren-Caldwell (1977)
Hydrophobic interaction	Formation of contacts between apolar groups in *aqueous solution*	Various proteins and nucleic acids	Yon (1972)
Covalent (chemisorption)	Formation of covalent disulfide bonds that can be cleaved again under mild conditions	Thiol groups containing peptides and proteins, mercurated polynucleotides	Brocklehurst et al. (1973)

TABLE 2.2 Special Techniques Based on the Principle of Affinity Chromatography (Related Techniques)

Method	Principle	Application	References
Affinity partition	Partition of a substance in a liquid two-phase system; matrix-bound ligand is solved in one of these phases	Laboratory and scale-up preparations	Flanagan and Barondes (1975), Hustedt and Kula (1977).
High-performance liquid affinity chromatography	Application of affinity adsorbents in high-performance liquid chromatography	Rapid separation of substances, preferentially in analytic scale	Ohlson et al. (1978)
Affinity electrophoresis	Application of affinity gels in common electrophoresis	Analytic scale	Bøg-Hansen (1973)
Analytic application of immunotechniques	Antigen-antibody interactions	Immunoassays	Yalow and Berson (1959)
		Immunoelectrodes	Aizawa et al. (1980)

Affinity chromatography at subzero temperatures	Chromatography below 0°C in salt containing aqueous solutions or mixed solvents	Separation of enzymes applying the immobilized substrates as ligands; applicability limited	Balny and Douzou (1979)
Affinity elution	Biospecific elution of substances adsorbed on ion exchangers or other matrices using the free ligands	Limited to preparative scale	Scopes (1977a, 1977b, 1982)
Affinity electrode technique	Potential change of a specially constructed electrode by protein adsorption	Analytic scale	Lowe (1979); Lowe et al. (1982)
Affinity therapy	Adsorption of toxins, xenobiotics, and pathogenic immunoactive substances in vivo by extracorporeal devices	Therapeutic purposes in medicine	Ringsdorf (1975); Kálal et al. (1982)

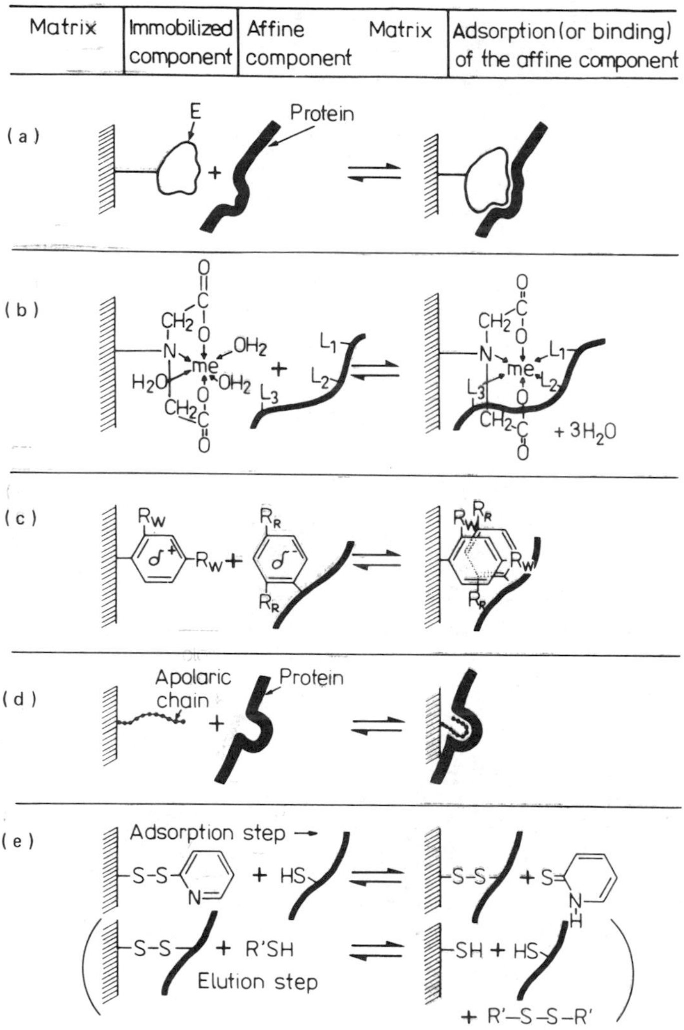

FIGURE 2.1 Variants of affinity chromatography. (a) Biospecific affinity chromatography (biosorption), (b) metal chelate chromatography, (c) charge transfer adsorption chromatography, (d) hydrophobic interaction chromatography (as proposed by Shaltiel et al., 1978), and (e) covalent chromatography (chemisorption). Abbreviations: E = enzyme, L = amino acid group, me = metal ion, R_W = electron-withdrawing substituent, R_R = electron-donating substituent. The thick lines symbolize the protein.

REFERENCES

Aizawa, M., Morioka, A., and Suzuki, S. (1980). An enzyme immuno sensor for the electrochemical determination of the tumor antigen α-fetoprotein. *Anal. Chem. Acta 115:* 61-67.

Axen, R., Porath, J., and Ernback, S. (1967). Chemical coupling of peptides and proteins to polysaccharides by means of cyanogen halides. *Nature 214*: 1302-1304.

Balny, C., and Douzou, P. (1979). Affinity chromatography at subzero temperatures. In *Affinity Chromatography and Molecular Interactions*, J.-M. Egly (ed.). Editions INSERM, Paris, pp. 99-107.

Bøg-Hansen, T. C. (1973). Immuno-affinoelectrophoresis; an analytical method to predict the result of affinity chromatography. *Anal. Biochem. 56*: 480-488.

Brocklehurst, K., Carlsson, J., Kierstan, M. P, J., and Crook, E. M. (1973). Covalent chromatography. Preparation of fully active papain from dried papaya latex. *Biochem. J. 133*: 573-584.

Campbell, D. H., Luescher, E., and Lerman, L. S. (1951). Immunologic adsorbents. I. Isolation of antibody by means of a cellulose protein antigen. *Proc. Nat. Acad. Sci. U.S. 37*: 575-578.

Čoupek, J., Křivaková, M., and Porkorný, S. (1973). New hydrophilic materials for chromatography: Glycol methacrylates. *J. Polym. Sci., Polym. Symp. No. 42*: 185-190.

Cuatrecasas, P., and Wilchek, M. (1968). Single-step purification of avidin from egg white by affinity chromatography on biocytin-Sepharose columns. *Biochem. Biophys. Res. Commun. 33*: 235-239.

Cuatrecasas, P., Wilchek, M., and Anfinsen, C. B. (1968). Selective enzyme purification by affinity chromatography. *Proc. Nat. Acad. Sci. U.S. 61*: 636-643.

Easterday, R. L., and Easterday, J. M. (1974). In *Immobilized Biochemicals and Affinity Chromatography*, R. B. Dunlap (ed.). Plenum, New York, p. 123.

Engelhardt, W. A., Kisselev, L. L., and Nezlin, R. S. (1970). Das Prinzip des fixierten Partners als Werkzeug auf dem Gebiet der Molekularbiologie. *Monatsh. Chem. 101*: 1510-1517.

Epton, R., Holloway, C., and McClaren, J. V. (1974). Cross-linked poly(acryloylmorpholinos) as matrices for gel permeation chromatography. *J. Appl. Polym. Sci. 18*: 179 192.

Flanagan, S. D., and Barondes, S. H. (1975). Affinity partitioning. A method for purification of proteins using specific polymer ligands in aqueous polymer two-phase systems. *J. Biol. Chem. 250*: 1484-1489.

Hayman, M. J., Skehel, J. J., and Crumpton, M. J. (1973). Purification of virus glycoproteins by affinity chromatography using lens culinaris phytohaemagglutinin. *FEBS Lett. 29*: 185-188.

Hjerten, S. (1962). Chromatographic separation according to size of macromolecules and cell particles on columns of agarose suspensions. *Arch. Biochem. Biophys. 99*: 466-475.

Hustedt, H., and Kula, M.-R. (1977). Studies of the interaction between aminoacyl tRNA synthetase and transfer ribonucleic acid by equilibrium partition. *Eur. J. Biochem. 74*: 191-198.

Kálal, J., Drobník, J., and Rypácek, F. (1982). Affinity chromatography and affinity therapy. In *Affinity Chromatography and Related Techniques*, T. C. J. Gribnau, J. Visser, and R. J. F. Nivard (eds.). Elsevier Scientific, Amsterdam, pp. 365-374.

Kristiansen, T. (1975). Virus purification by *Vicia ervilia* lectin coupled to Sepharose. *Prot. Biol. Fluids 23*: 663.

Larsson, P.-O., and Mosbach, K. (1977). Affinity precipitation of enzymes. *FEBS Lett. 98*: 333-338.

Lowe, C. R. (1979). The affinity electrode; application to the assay of human serum albumin. *FEBS Lett. 106*: 405-408.

Lowe, C. R., Clonis, Y. D., Goldfinch, M. J., Small, D. A. P., and Atkinson, A. (1982). Some preparative and analytical applications of triazine dyes. In *Affinity Chromatography and Related Techniques*, T. C. J. Gribnau, J. Visser, and R. J. F. Nivard (eds.). Elsevier Scientific, Amsterdam, pp. 389-398.

Mosbach, K., and Andersson, L. (1977). Magnetic ferrofluids for preparation of magnetic polymers and their application in affinity chromatography. *Nature 270*: 259-261.

Ohlson, S., Hansson, L., Larsson, P. O., and Mosbach, K. (1978). High performance liquid affinity chromatography (HPLAC) and its application to the separation of enzymes and antigens. *FEBS Lett. 93*: 5-9.

Porath, J. (1978). Explorations into the field of charge transfer adsorption chromatography. *J. Chromatogr. 159*: 13-24.

Porath, J. (1982). Affinity chromatography – historical survey – present status – future aspects. In *Affinity Chromatography and Related Techniques*, T. C. J. Gribnau, J. Visser, and R. J. F. Nivard (eds.). Elsevier, Amsterdam, pp. 3-8.

Porath, J., and Dahlgren-Caldwell, K. (1977). Charge-transfer adsorption chromatography. *J. Chromatogr. 133*: 180-183.

Porath, J., and Flodin, P. (1959). Gel filtration: A method for desalting and group preparation. *Nature 183*: 1557-1559.

Ringsdorf, H. (1975). Structure and properties of pharmacologically active polymers. *J. Polym. Sci., Polym. Symp. No. 51*: 135-153.

Scopes, R. K. (1977a). Purification of glycolytic enzymes by using affinity elution chromatography. *Biochem. J. 161*: 253-263.

Scopes, R. K. (1977b). Multiple enzyme purification from muscle extracts by using affinity elution chromatographic procedures. *Biochem. J. 161*: 265-277.

Scopes, R. K. (1982). Affinity elution from ion exchangers principles; problems and practice. In *Affinity Chromatography and Related Techniques*, T. C. J. Gribnau, J. Visser, and R. J. F. Nivard (eds.). Elsevier, Amsterdam, pp. 333-339.

Shaltiel, S., Halperin, G., Er-el, Z., Tauber-Finkelstein, M., and Amsterdam, A. (1978). Homologous series of hydrocarbon coated agarose in hydrophobic chromatography. In *Affinity Chromatography*, O. Hoffmann-Ostenhof, M. Brietenbach, F. Koller, D. Kraft, and O. Scheiner (eds.). Pergamon Press, Oxford, pp. 141-160.

Sharon, N. (1979). Application of lectins to cell fractionation. In *Affinity Chromatography and Molecular Interactions*, J.-M. Egly (ed.). Editions INSERM, Paris, pp. 197-205.

Starkenstein, E. (1910). Über Fermentwirkung und deren Beeinflussung durch Neutralsalze. *Biochem. Z. 24*: 210-218.

Yalow, R. S., and Berson, S. A. (1959). Assay of plasma insulin in human subjects by immunological methods. *Nature 184*: 1648-1649.

Yon, R. J. (1972). Chromatography of lipophilic proteins on adsorbents containing mixed hydrophobic and ionic groups. *Biochem. J. 126*: 765-767.

part II
GENERAL PROBLEMS

3
Solid Matrix Support

An essential prerequisite for affinity chromatography is the availability of appropriate chromatographic matrices with the covalently bound specific ligand. Unlike the classic chromatographic procedures, which require only a few basic types of matrices, such as dextrans with varying degrees of cross-linking or molecular sieves or ion exchangers of different charge or ionic strength, a matrix for affinity chromatography, with regard to the substances to be separated or purified, is adapted to the given purpose by the correspondingly selected ligand. In many cases the matrix can be used for a specific purification step only; that is, it is tailored to the purification of the corresponding substance.

The working steps of affinity chromatography – adsorption, washing, and elution – are mostly simple to perform without major problems. The time-consuming and frequently limiting step is the search for an appropriate complex partner and the preparation of the bioaffine matrix.

At present a variety of bioaffine matrices is commercially available as so-called ready-to-use adsorbents. In many cases it is necessary and appropriate to have a homemade matrix for affinity chromatography. The synthesis of a matrix for affinity chromatography requires

1. A matrix support suitable for chromatography
2. A substance with specific or selective affinity to the substance to be purified
3. A chemical reaction for covalent linkage of the two partners

Therefore, the synthesis and properties of an affinity chromatographic matrix will be discussed in more detail.

TABLE 3.1 Some Specific Properties of Supports for the Affinity Chromatography

Hydrophilic	but	insoluble in water
Macroporous	but	mechanically stable
Chemically stable	but	easily chemical modified
Great surface	but	inert, not nonspecific adsorption

GENERAL PROPERTIES OF THE SOLID MATRIX SUPPORT

Up to now, a variety of matrix supports has been used with more or less success for affinity chromatography. The correct choice of matrix support and the covalent linkage between the matrix and the bioaffine ligand may be essential for the success of the chromatography.

As is shown in the following chapters, the supporting matrix plays not only a passive role as the solid-phase component; it can also have a considerable effect on the stability of the complex formed, or it may even be the bioaffine ligand itself. Furthermore, the manner and site of linking with the bioaffine ligand may have an essential influence on the effectivity of affinity chromatography (Chap. 5).

The matrix should have properties generally required for a chromatographic matrix and, furthermore, qualities that derive from the specificity of the affinity chromatography. The general properties include adequate particle size and shape and sufficient stability and surface. Increasing particle size reduces flow resistance and separation power. On the other hand, matrices with very low particle size in consequence have too high a flow resistance and soon become clogged. Irregularly shaped particles lead to unequal path lengths for the substances to be separated and, consequently, to band broadening. Best suited is the spherical shape.

Also important are mechanical and chemical stability and resistance against microorganisms. A large surface is desirable, which can be reached best by using a porous and swellable material with a highest possible ratio of the inner to the outer surface. It is not possible here to go into further detail; the reader is referred to the standard handbooks and monographs on chromatography (Lederer and Lederer 1957; Determann, 1969; Krauss and Krauss, 1981).

A matrix for affinity chromatography should possess several specific properties, some of which appear contradictory but are associated with the nature of the substances to be purified and with the kind of interactions to be utilized (Table 3.1).

The aqueous milieu is an essential condition for structure and/or function of biologically active compounds. In some cases water molecules are a part of the native configuration of a biomacromolecule. This is why affinity chromatography is generally performed in the aqueous phase. Thus, the matrix must not only be water insoluble but also hydrophilic, swellable, or, at least, well wettable with water (the matrix itself or at least its surface).

The matrix should be either macroporous or consist of a wide-mesh network to allow free diffusion of the biomacromolecules (proteins, enzymes, nucleic acids, antigens, and others) to the binding sites and not prevent biospecific complex formation by a molecular sieve effect. High porosity or a wide-mesh network generally decreases the mechanical stability of the matrix material.

Furthermore, the matrix should be chemically stable in a wide range (pH 2-12). On the other hand, chemical modification – incorporation of functional groups for covalent binding of the complex partner (Chap. 4) – must be possible in a simple way.

The supporting material should have the greatest possible surface and at the same time be completely inert and have no unspecific interactions with the substances to be purified, such as electrostatic or hydrophobic interactions. But depending on the nature of the matrix, these never can be completely excluded and have been observed increasingly after covalent fixation of the bioaffine ligand either through the incorporation of dissociable groups as a result of functionalization or hydrophobic spacers (Chap. 4). This must not necessarily be a disadvantage. These "nonspecific interactions" may amplify, for example, complex formation. As hydrophobic chromatography shows, such interactions may in the extreme case lead to a new principle of separation (Chap. 15).

Up to the present time, there is no supporting matrix that has all the requirements in an optimal way. Thus, various supporting materials have been used more or less successfully (Table 3.2). Within the scope of this chapter only a few basic types and their advantages and limitations can be described and briefly discussed.

BIOPOLYMERS

Hydrophilic biopolymers play a dominating role as supporting materials for affinity chromatography. Starting materials are natural polysaccharides, such as agarose, dextran, cellulose, and, to a lesser extent, starch. The modification necessary for affinity chromatography can be carried out in a relatively simple way via the OH groups (Chap. 4). In spite of many common features, there are distinct differences in properties and applications relating to the chemical structure of the biopolymers (Fig. 3.1). This is discussed in detail by Madden and Thom

TABLE 3.2 Some Solid Matrix Supports for the Affinity Chromatography

Type	Chemical structure
Biopolymers (polysaccharides)	Agarose (cross-linked, beads, macrobeads); cross-linked dextran; cellulose (microcrystalline, macroporous)
Synthetic copolymers	Polyacrylamide; poly(hydroxyethyl)-methacrylate; polystyrene
Inorganic material	Porous glass, iron oxide (magnetogels)
Biopolymers/synthetic copolymers	Agarose polyacrylamide
Inorganic materials/organic copolymers	Silica/hydrophilic copolymers

(1982). The conformation of the chains and the interaction between them have an influence on the use of polysaccharides as chromatographic support. The geometry of the chains is determined by the relative orientation of adjacent sugar residues around the glycosidic bonds. But in aqueous solutions a disordered random coil form dominates, because of the flexibility of the chains. Under definable conditions, however, non-bonded energy-terms (hydrogen bridges, ionic interactions, and others) can compensate for conformation entropy and fix the polysaccharide chains in an ordered configuration. Of great practical importance for the optimization of polysaccharides as chromatographic support is that it is possible to influence the factors responsible for the equilibrium between an ordered and a non-ordered configuration in aqueous solution.

On the basis of their covalent structure, the polysaccharides can be divided into two classes:

1. Simple periodic structures
2. Interrupted periodic structures

In the first class the chains contain only identical sugar residues, which are linked through identical positions and configurations. Four types of conformation are known:

Extended ribbons, with cellulose
Coiled springs, with starch

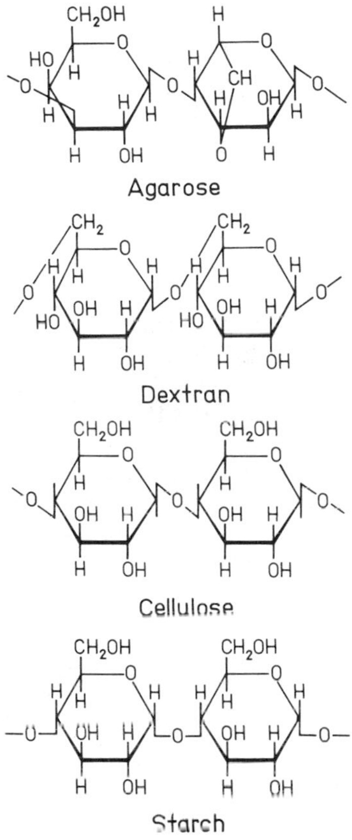

Agarose

Dextran

Cellulose

Starch

FIGURE 3.1 Partial structure of some common polysaccharides.

Crumpled ribbons
Flexible coils, with dextran as representative

The second class is characterized by chains containing periodic sequences with ordered conformations, which are interrupted by deviations from regularity. The balance between ordered and "soluble" regions leads to highly hydratized gels with manyfold applications. Agarose is representative of the interrupted periodic type. Other types, such as alginates and pectins, with interrupted ribbon configurations are of great interest in the application of immobilized whole cells but not for chromatographic supports.

Agarose

Agarose was introduced into affinity chromatography as a matrix material by Cuatrecasas and colleagues (1968). This classic work in connection with the work done by Porath and his group on cyanogen bromide as a coupling agent (Axen et al., 1967; Porath et al., 1967) led to the breakthrough of this method. Supporting materials based on agarose have been the matrices of choice until now, above all for laboratory use. Along with agaropectin, agarose is a component of agar, which may be isolated from different species of the Rhodophyceae family of red seawater algae (Guiseley and Renn, 1977).

The separation of agarose from the polysaccharide mixture is based on differences in solubility and/or chemical reactivity, which is associated with the anionic character of agaropectins. Agarose is a linear water-soluble polysaccharide composed of alternating 1,3-linked β-D-galactose and 1,4-linked 3,6-anhydro-α-L-galactose units (Fig. 3.1).

That aqueous agarose solutions gel spontaneously when cooled below 50°C has led to the development of agarose gels in bead, pellet, or spherical forms. Practical preparation methods have been described by Hjerten (1964), as well as Bengtsson and Philipson (1964). The excellent properties as chromatographic supporting material have been attributed to the particular structure of the gel (Fig. 3.2).

From x-ray studies it may be postulated (Rees, 1972; Arnott et al., 1974; Madden and Thom, 1982) that the polysaccharide chains form a double helix, then aggregate via hydrogen bridges and hydrophobic interaction into fibers or bundles with ordered structures. The gel structure is maintained by "junction zones," including noncovalent linkages. This network phase may contain up to 100 parts of water per polysaccharide moiety and leads to the relatively large spaces through which the biomacromolecules can diffuse. The exclusion limit may be varied within wide ranges because the pore size is inversely proportional to the agarose concentration (Table 3.3). Porath (1981) holds the view that adsorption caused by bound water acts as a screen that keeps the solutes away from the matrix and thus provides the biopolymers with a mild environment.

The agarose matrices are very hydrophilic, mechanically relatively stable, readily modifiable chemically via the OH groups, and normally do not have unspecific properties. Nevertheless, under certain conditions the weak unspecific interactions become distinctly evident. They are attributable, on the one hand, to a residual content of sulfate groups (below 0.7%) in the form of the ester in position 6 (CH_2-OSO_3^-) or carboxyl group and, on the other hand, to hydrophobic interactions, probably attributed to the ether bridge of the anhydrogalactose unit. At high salt concentrations the latter leads to the adsorption of proteins and nucleic acids (Holmes et al., 1975; Zeichner and Stern, 1977).

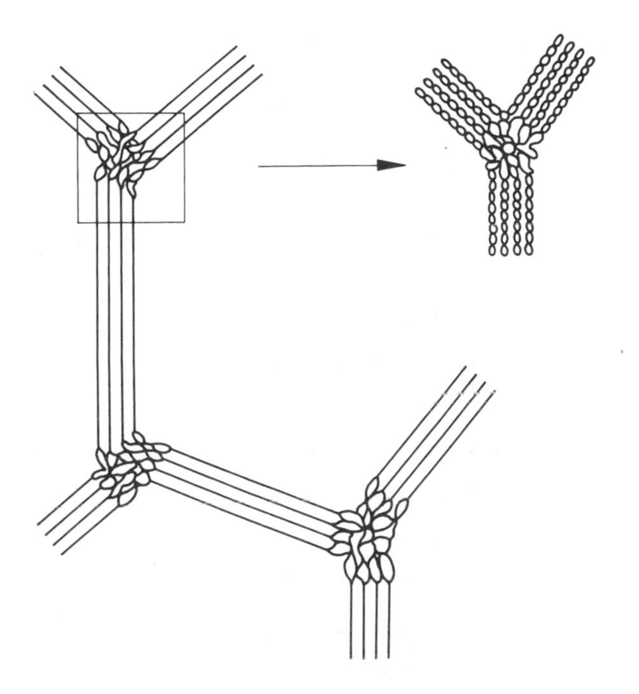

FIGURE 3.2 Partial structure of agarose gel.

Agarose matrices for gel filtration and affinity chromatography are offered by various producers, mainly by Pharmacia (Uppsala, Sweden) under the trade name Sepharose, by Bio Rad (Richmond, California) under the trade name Bio-Gel A, or by Reactifs IBF (Paris) under the trade name Ultrogel A distributed worldwide by LKB-Producter AB (Bromma, Sweden). Several types are listed in Table 3.3.

The various types are distinguished mainly by the exclusion volume, which may vary from 0.5×10^6 (10% agarose) to 150×10^6 (1% agarose) solely by altering agarose concentration.

Agarose is relatively stable over a wide range of experimental conditions, as shown in Table 3.4 for Sepharose. It also has considerable disadvantages. It cannot be heat sterilized, and one cannot work at elevated temperatures or with organic solvents, which is frequently necessary for modification to couple bioaffine ligands.

The stability of the agarose matrix against mechanical, chemical, or microbiological degradation can be considerably increased by chemical cross-linking (Table 3.4). For this purpose, various procedures with low-molecular-weight bifunctional agents have been developed, for

TABLE 3.3 Some Commercial Types of Agarose Gels

Trade name	%[a]	Fractionation range (molecular weight)
Bio-Gel A-0.5 m	10	1×10^4 to 5×10^5
Bio-Gel A-1.5 m	8	1×10^4 to 1.5×10^6
Bio-Gel A-5 m	6	1×10^4 to 5×10^6
Sepharose 6B/Cl-6B	6	1×10^4 to 4×10^6
Ultrogel A6	6	2.5×10^4 to 4×10^6
Bio-Gel A-15 m	4	4×10^4 to 1.5×10^7
Sepharose 4B/Cl-4B	4	6×10^4 to 2×10^7
Ultrogel A4	4	5.5×10^4 to 2×10^7
Bio-Gel A-50 m	2	1×10^5 to 5×10^7
Sepharose 2B/Cl-2B	2	7×10^4 to 4×10^7
Ultrogel A2	2	1.2×10^5 to 5×10^7
Bio-Gel A-150 m	1	1×10^6 to 1.5×10^8

[a]Agarose in gel.

example, with epichlorohydrin (Porath et al., 1971; Kristiansen, 1974), 2,3-dibromopropanol (Kristiansen, 1974; Lååas, 1975), and divinylsulfone (Kristiansen, 1974; Porath et al., 1975). Other soluble polysaccharides can also be cross-linked (see the next section). Young and Leon (1978), for example, described the cross-linkage of guarum (guar gum), a soluble galactomannan, and dextran with divinyl sulfone and their application to the purification of lectins from *B. simplicifolia* and *A. hypogaea*.

The general instability of polysaccharides under oxidizing conditions may also occur after cross-linking. The structure of cross-linked agarose is analogous to that of Sephadex. However, cross-linking of Sepharose does not decrease the effective pore size, thus suggesting that cross-links take place mainly between chains in a single gel fibers, probably between oxygen in position 6 of the galactose residue and oxygen in position 2 of the anhydrogalactose residue (Madden and Thom, 1982). The Sepharose CL matrices are important especially as starting material for matrices when drastic conditions are required for the modification or direct conduct of chromatography. It

TABLE 3.4 Comparison of Stability Between Normal and Covalent Cross-Linked Agarose Gels

	Sepharose	Sepharose CL
pH	4-9	3-14
Temperature (°C)	0-40	<70
Solvents	Aqueous solutions containing high concentrations of salts, urea, guanidine·HCl, detergents	Aqueous solutions, 6 M guanidine·HCl, 8 M urea and detergents in a pH range 3-11
	Dimethylformamide-H_2O (1:1)	Organic solvents
	Ethyleneglycol-H_2O (1:1)	
Sterilization	Not autoclaving	Autoclaving at pH 7 and 110-120°C
Chaotropic ions (KSCN)	Low stability	High stability

must be considered that, as a result of cross linking, the number of activatable OH groups is distinctly lower than with non-cross-linked gels. Various derivatives of agarose modified especially for application as affinity resin are offered commercially.

Supports with large beads are advantageous for the affinity chromatography of cells, so that they can pass through the column without being physically trapped (Chap. 10). For this purpose an agarose gel has been developed with the trade name Sepharose 6 MB. The macrobeads have a large diameter (250-350 μm), uniform shape, and low nonspecific adsorption of cells.

Agarose is also used in combination with synthetic polymers as a supporting substance. The Ultrogel AcA (LKB-Producter AB, Bromma, Sweden), for example, consists of a three-dimensional polyacrylamide lattice and an interstitial agarose gel of a composition different from that of both partners. The acrylamide-agarose ratio may vary from 1:1 to 10:1.

Recently, Porath (1981) has described a new type of agar-agarose gel that combines the advantages of mechanical stability with high permeability and effective complex formation. At first, small particles (silica) or macromolecules are mixed with the agarose gel. Then the gel is shrunk by substituting methanol for water and is cross-linked in this state. Subsequently, the entrapped particles or macromolecules are released again. Analogous developments have been described with other types of matrices (see the last section). The basic idea of these developments is to coat the surface of a mechanically stable and porous supporting material not suitable for affinity chromatography because of its high nonspecific interaction, or poor faculty to undergo chemical modification, with a layer that is hydrophilic, readily modifiable chemically, and shows no nonspecific interaction.

Cross-linked Dextran

Dextran is a polysaccharide produced by microorganisms of genus *Leuconostos*. It is composed of 1,6-linked α-D-glucose units (over 90%) and can be branched out by 1,2, 1,3, or 1,4 binding (Fig. 3.1). Dextran cross-linking with epichlorohydrin in alkaline solution is one of the classic supports for gel permeation chromatography. Depending on chain length of the dextran and the degree of cross-linking, gels with well-defined molecular screening effects can be produced. Gels with a low degree of cross-linking, that is, with a high exclusion limit, have been used with good success in affinity chromatography. The main representative of this type is produced by Pharmacia (Uppsala, Sweden) under the trade name Sephadex. Other commercial matrices on this basis are Molselect (Reanal, Hungary) and Epidex (VEB Serumwerk, Bernburg, GDR). Reviews on cross-linked dextran have been given by Porath (1962) and Kennedy (1974).

Sephadex is a gel derived from a three-dimensional network in bead form. Its partial structure is given in Figure 3.3. The strongly hydrophilic character is due to the high content of OH groups. The Sephadex matrices are insoluble in all solvents, provided they are not chemically degraded. They are stable in water, salt solutions, organic media, and even in alkaline solutions. Strong acids, however, split the glycoside links, but they can be treated for 1-2 hr with 0.1 M HCl without appreciable degradation. In 0.02 M HCl they are stable for 5 months.

In aqueous solutions the swelling behavior is independent of pH and ionic strength. Also, the swelling and drying processes are completely reversible. Oxidants, however, cause degradation of the matrix.

Recently, a molecular sieve has been developed by Pharmacia (Uppsala, Sweden) under the trade name Sephacryl (Johansson and Lundberg, 1979). The sieve is produced by covalent cross-linking of allyl-dextran with N,N'-methylene-bis-acrylamide. The big advantage of this kind of matrix is the excellent flow rate, because this support is exceptionally rigid but the unspecific adsorption is increased.

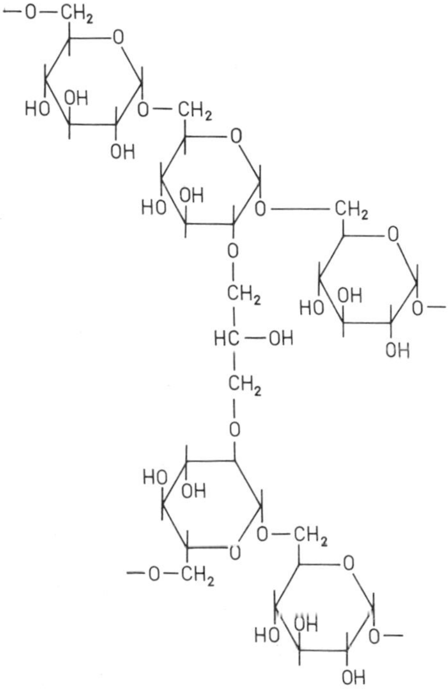

FIGURE 3.3 Partial structure of cross-linked dextran (Sephadex).

Cellulose

Since the introduction of chromatographic procedures, cellulose has been one of the standard supporting materials. At first, it was also prevalent in affinity chromatography, as is evident from the first studies in the 1950s, such as that by Campbell and coworkers (1951), who isolated bovine albumin antiserum of rabbits on bovine albumin, and Lerman (1953), mushroom tyrosinase on phenylazophenole, both covalently bound to p-aminobenzylcellulose. Arsenis and McCormick (1964, 1966) used flavin derivatives of carboxymethyl/cellulose (CMC) for the purification of flavokinase and riboflavin 5'-phosphate (flavin mononucleotide; FMN)-dependent enzymes, respectively. Especially important was cellulose as the immunoadsorbent matrix (Weliky and Weetall, 1965).

Since the introduction of agarose, cellulose no longer has this importance, in spite of its low price and other advantages (stability). The reasons for this should be sought in its physical and chemical structure. Cellulose, a vegetable polysaccharide, is composed of linear 1,4-β-D glucose units (Fig. 3.1). The polysaccharide chains are aggregated to fibers with a high degree of crystallinity and microcrystalline regions separated by amorphous regions. The high degree of aggregation and molecular orientation is caused by hydrogen bridges. Cellulose is relatively stable against physical and chemical influences. The glycosidic bonds are hydrolyzable only under extremely acidic conditions. The cellulose swells in strongly alkaline solution (30-50%), as well as in pyridine, glacial acetic acid, and dimethylsulfoxide. This swelling process leads to an increase of the amorphous moiety. In general, the crystalline-amorphous ratio is extremely milieu dependent. Chemical modifications occur preferentially in the amorphous regions so that heterogeneity of the ligand density may be the result, which is frequently undesirable. Furthermore, the fibrous structure, which lacks uniformity, the unadequate geometric shape, and the absence of macroporosity prevent good penetration of macromolecules and lead to a relatively low ligand density and flow rate (Knight, 1967; Madden and Thom, 1982). These drawbacks can be overcome by two ways.

A cellulose-based matrix for affinity chromatography has been developed by Merck (Darmstadt, FRG). The high degree of substitution of the starting material CMC ensures that the chains are completely in the flexible coil form. Via methyl ester and the hydrazide, the CMC is converted into the azide, which is subsequently reacted with an excess of α,ω-diaminoalkanes (preferentially C_6 and C_{12}). In addition to the introduction of the aminoalkyl group, cross-linking is achieved simultaneously by the bifunctional reagent (Brümmer, 1974). On the basis of this aminoalkylcellulose, further modifications may be carried out. This type of carrier shows very good properties for affinity chromatography and is asserted to be equivalent to agarose. With 800 μmol ligand per gram matrix, or 80 μmol/ml, the capacity is five times that of agarose.

The second way is the use of macroporous reconstituted cellulose in beaded form. Various methods have been devised for the preparation of macroporous cellulose (Peška et al., 1976; Kuga, 1980). Štamberg and colleagues (1982) give a survey of the individual methods and in detail describe the bead cellulose they developed.

At present this type of support seems to be a real alternative to the agarose gels. The starting material is cellulose xanthogenate, and the important step of preparation – the solidification of liquid spherical droplets – is performed by thermal sol-gel transition.

The properties of the support prepared in this way are well suited for affinity chromatography (Chap. 12). They are regular, spherical beads of high porosity. The content of regenerated cellulose in the swollen beads is ∿15%, and the pore volume is 90% (calculated from the content of H_2O). The bead cellulose is mechanically more stable than the corresponding dextran or agarose gels. Bead cellulose can be used with good results as an agarose gel, although there are some differences. In its macroporous structure the cellulose shows a nonspecific interaction strongly dependent on the milieu. The exclusion limit is above 450,000 (the molecular weight of ferritin) and below 2,000,000 (the molecular weight of dextran blue).

Starch

The first examples of affinity chromatography were made with insoluble starch (Fig. 3.1) without recognizing the importance of this method, such as the complex formation between α-amylase and insoluble starch (Starkenstein, 1910) or the separation of α- and β-amylase (Holmbergh, 1933). Today, starch plays a secondary role as matrix, for example, as a derivative of dialdehyde starch (Goldstein, 1970; Goldstein et al., 1970). For this purpose, the starch is converted to dialdehyde derivative by periodate oxidation (Chap. 4) and then cross-linked with 4,4'-diaminodiphenylmethane. Fixation of the ligand can then be accomplished after reduction of the Schiff's base via diazotization.

SYNTHETIC POLYMERS

Besides supports based on biomacromolecules, synthetic polymers have been explored as a potential supporting material for affinity chromatography. This development proceeded in close connection with the investigation of immobilized enzymes. To date, a variety of materials of varying composition has been described but none have attained the universal application of agarose. In most cases they are cross-linked macroporous vinyl polymers in beaded or spherical form with defined size and porosity. The properties can be obtained by proper choice of monomers and conditions of polymerization. The chemical structure of these supports (1) is characterized by the polyethylene backbone, which influences chemical stability and physical or structural stability, and the modifiable side chains R_1, R_2, R_3, R_4:

$$-\underset{R_3}{\overset{R_1}{\underset{|}{\overset{|}{C}}}}-\underset{R_4}{\overset{R_2}{\underset{|}{\overset{|}{C}}}}- \qquad (1)$$

Only the most important representatives of the synthetic polymers can be characterized here.

Polyacrylamide Supports

Polyacrylamide gels are copolymerizates between acrylamide and N,N'-methylene-bis-acrylamide as cross-linking agents containing a hydrocarbon framework with carboxamide side chains:

$$-CH_2-\underset{\displaystyle CO-NH_2}{\underset{|}{CH}}-CH_2-\underset{\displaystyle CO-NH_2}{\underset{|}{CH}}- \qquad (2)$$

Polyacrylamide gels mainly are offered by Bio-Rad (Richmond, California) under the trade name Bio-Gel-P as a carrier for gel permeation, which then can be functionalized for affinity chromatography.

This support can be prepared with a highly varying range of pore sizes; the exclusion limit can differ between 1800 (Bio-Gel P-2) and 400,000 (Bio-Gel P-300). It is hydrophilic, stable in the pH range 1-10, biologically inert, and not attached by microorganisms and shows only minimal properties of ion exchange based on carboxamide side chains. The high content of these carboxamide side chains is advantageous for affinity chromatography, because the functionalization of these groups allowed the covalent binding of ligands. Inman and Dintzis (1969) have developed methods for the modification of these groups by means of polymer analog reactions under mild conditions (Chap. 4). Other modified polyacrylamides for affinity chromatography are supports bearing the trade name Enzacryl (Koch-Light, Colnbrook, UK). In contrast to the Bio-Gel matrices, these are synthesized by copolymerization of the modified monomers (Epton and Thomas, 1971).

Hydroxyalkylmethacrylate Supports

The well-known matrices of this type are those from Czechoslovakia marketed under the trade name Spheron (Lachema Brno) and Separon (Laboratory Instruments Works). In the early 1970s this type was synthesized by Čoupek and coworkers (Čoupek et al., 1973) and introduced to affinity chromatography by the investigations of Turková (Turková et al., 1973; Turková and Seifertová, 1978). Spheron and Separon are prepared by heterogeneous suspension copolymerization of hydroxyethylmethacrylates and ethylenedimethacrylates in aqueous solution in the presence of inert solvents. This gives a neutral, hydrophilic gel from heavily cross-linked microparticles with micropores, aggregating to macroparticles with macroporous structure (Fig. 3.4). The inner structure, pore size, and distribution, specific surface, and quantity of reactive OH groups can be varied, with a molecular weight exclusion limit from 20,000 to 20,000,000.

In contrast to other hydrophilic types of matrix, the macroporous structure is kept in dry state. The gel does not change in volume with changes of pH or in organic solvents and is not attacked by microorganisms as are other synthetic polymers. The gel can be used with good results in organic solvents. This is a great advantage for the modification of the matrix by polymer-analogous reactions. The mechanical and chemical stability is higher than that of matrices based on acrylic acid derivatives. Because of their high rigidity this matrix

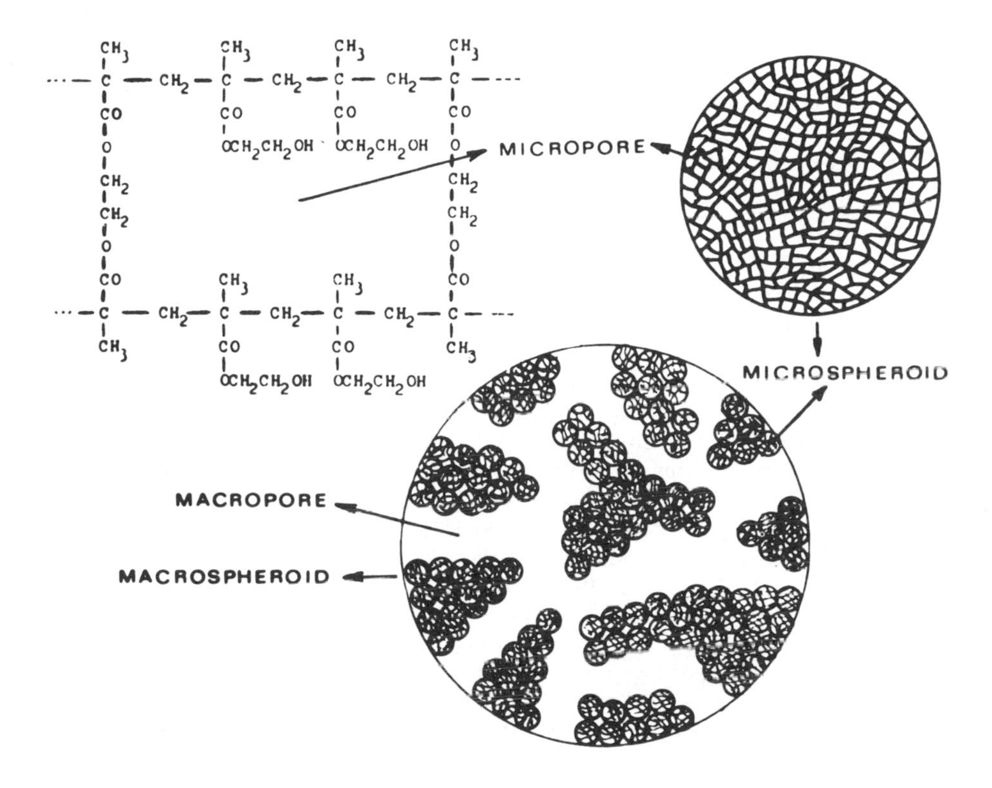

FIGURE 3.4 Structure of hydroxyalkylmethacrylate gel (Spheron, Separon). (From Turková and Seifertová, 1978, *J. Chromatogr. 148*: 293-297)

shows excellent flow properties, and Separon is a matrix for high-performance liquid chromatography. Thermic decomposition is observed only above 250°C. The CH_3 group at the α C-atom leads to clear protection of the ester group against hydrolysis because of steric factors.

The matrix is stable after heating in 1 M Na glycolate for 8 hr or in 20% HCl for 24 hr at 150°C. These properties allowed the application of this support to large-scale operations and to industrial production.

The chemically reactive groups for the covalent coupling of biospecific ligands are the OH groups of the side chains. In comparison with the agarose, the hydroxyethylmethacrylates have a greater nonspecific adsorption, attributable largely to hydrophobic interactions. This hydrophobic property can be used for hydrophobic interaction chromatography of proteins and peptides (Štrop et al., 1978).

Other methacrylate gels containing reactive groups are also suitable carriers for use in affinity chromatography, for example, p-nitrophenylester (Turková, 1976; Čoupek et al., 1977) or epoxide (oxiran) groups — copolymers of glycidyl methacrylate (Švec et al., 1975, 1977; Turková et al., 1978).

Other Supports

Another group of supports contains anhydride residues as the active center for covalent attachment; these are copolymers of ethylene and maleic anhydride (EMA) (Goldstein, 1970, 1972), maleic anhydride and butane-diol-divinyl ether (Brümmer et al., 1972), or acrylamide and methacrylic anhydride (Krämer et al., 1974 a, b).

These supports have a high capacity for ligand binding but are suitable only to a limited extent. After coupling of ligands, free carboxylic residues appear and impart to the matrix an ion-exchange character.

Polystyrene derivatives were used initially as a matrix for immunosorbents and are used today in enzyme-linked immunoadsorbent assay kits (ELISA) (Engvall and Perlmann, 1971). This material is not suitable for broad application in affinity chromatography because of the high hydrophobicity, which can denature the adsorbed or covalently fixed protein. More suitable supports have been developed by Manecke and coworkers (Manecke, 1964) on the basis of copolymers of methacrylic acid and methacrylic acid fluoroanilide.

INORGANIC SUPPORTS

Like cellulose, inorganic materials are usually applied to many chromatographic methods (adsorption, distribution, gas chromatography, and others). Their advantages include high mechanical and chemical stability (particle and pore size) and resistance against microorganisms. At present, inorganic matrices are predominantly used in the field of immobilized enzymes, but as a result of the studies by Weetall and coworkers it has been possible to use porous glass as a matrix for affinity chromatography (Weetall, 1973; Weetall and Filbert, 1974).

Controlled-pore glass (CPG) consists of 97% SiO_2 and 3% B_2O_3. Its production is based on the following principle. Certain borosilicate glasses separate during heat treatment (annealing; 500-700°C) into two phases. One phase is nearly pure silica and insoluble in acids. The other phase is nearly pure sodium borate, pervades the other phase like an arterial network, and can be leached out by acid treatment. Pore diameters ranging from about 30 to 3000 Å with narrow pore size distribution can be prepared by controlling various parameters. The large surface of the porous glass leads to several undesirable properties: the nonspecific adsorption (Cuatrecasas and

TABLE 3.5 Some Functional Monomers for Coating

Allylamine
N-Allyl-1-bromoacetamide
6-Acrylamidohexanoic acid
N-Hydroxysuccinimidyl-6-acrylaminohexanoic acid ester
N-Hydroxysuccinimidylacrylic acid ester

Anfinsen, 1971) and the relatively high solubility of silica at the surface. These two effects can be reduced by surface treatment with hydrophilic silane, such as γ-aminopropyltriethoxysilane (Weetall and Filbert, 1974). Another possibility is glyceropropylsilylation (Regnier and Noel, 1976). This silylation to an aminoalkyl glass is the first step necessary for the covalent fixation of affine ligands (see Chap. 19). A special type of inorganic support is the "magneto-gels." A permanent magnet, such as iron oxide particles (magnetide), could be loaded directly with a polysaccharide, protein, or enzyme (Horisberger, 1976) or could be silaninized or entrapped in a gel (Mosbach and Andersson, 1977). Analogously with the immobilized enzymes they have the advantage of being simply separable from colloidal suspensions or undissolved materials. Not all properties of this type are suitable for use in affinity chromatography.

Schutyser and coworkers (1982) have shown another way of synthesizing new support materials, particularly for industrial application. By coating the surface of macroporous silica with a thin skin of hydrophilic copolymer containing spacer and/or functional groups, the advantages of both partners are combined to a tailor-made support. Some monomers are listed in Table 3.5

This support material shows an excellent flow rate, is completely inert in common organic solvents and aqueous solution between pH 2 and 9, and is regenerable by pyrolysis to blank silica. It can be used for the purification of heparin on a semi-industrial scale.

REFERENCES

Arnott, S., Fulmer, A., Scott, W. E., Dea, I. C. M., Moorhouse, R., and Rees, D. A. (1974). The agarose double helix and its function in agarose gel structure. *J. Mol. Biol. 90*: 269-284.

Arsenis, C., and McCormick, D. B. (1964). Purification of liver flavokinase by column chromatography on flavin-cellulose compounds. *J. Biol. Chem. 239*: 3093-3097.

Arsenis, C., and McCormick, D. B. (1966). Purification of flavin mononucleotide-dependent enzymes by column chromatography on flavin phosphate cellulose compounds. *J. Biol. Chem. 241*: 330-334.

Axen, R., Porath, J., and Ernback, S. (1967). Chemical coupling of peptides and proteins to polysaccharides by means of cyanogen halides. *Nature 214*: 1302-1304.

Bengtsson, S., and Philipson, L. (1964). Chromatography of animal viruses on pearl-condensed agar. *Biochim. Biophys. Acta 79*: 399-406.

Brümmer, W. (1974). Affinitätschromatographie. *Kontakte (Merck) 1/74*: 23-29; *2/74*: 3-13.

Brümmer, W., Hennrich, N., Klockow, M., Lang, H., and Orth, H. D. (1972). Preparation and properties of carrier-bound enzymes. *Eur. J. Biochem. 25*: 129-135.

Campbell, D. H., Luescher, E., and Lerman, L. S. (1951). Immunologic adsorbents. I. Isolation of antibody by means of a cellulose protein antigen. *Proc. Nat. Acad. Sci. U. S. 37*: 575-578.

Čoupek, J., Křiváková, M., and Pokorný, S. (1973). New hydrophilic materials for chromatography: Glycol methacrylates. *J. Polym. Sci., Polym. Symp. 42*: 185-190.

Čoupek, J., Labský, J., Kálal, J., Turková, J., and Valentová, O. (1977). Reactive carriers of immobilized compounds. *Biochim. Biophys. Acta 481*: 289-296.

Cuatrecasas, P., and Anfinsen, C. B. (1971). Affinity chromatography. *Annu. Rev. Biochem. 40*: 259-275.

Cuatrecasas, P., Wilchek, M., and Anfinsen, C. B. (1968). Selective enzyme purification by affinity chromatography. *Proc. Nat. Acad. Sci. U. S. 61*: 636-643.

Determann, H. (1969). *Gel Chromatography*. Springer Verlag, Berlin.

Engvall, E., and Perlmann, P. (1971). Enzyme-linked immunosorbent assay, ELISA. III. Quantitation of specific antibodies by enzyme-labeled anti-immunoglobulin in antigen-coated tubes. *J. Immunol. 109*: 129-135.

Epton, R., and Thomas, T. H. (1971). Improving nature's catalysts. *Aldrichim. Acta 4*: 61-65.

Goldstein, L. (1970). Water-insoluble derivatives of proteolytic enzymes. *Methods Enzymol. 19*: 935-962.

Goldstein, L. (1972). Microenvironmental effects on enzyme catalysis. A kinetic study of polyanionic and polycationic derivatives of chymotrypsin. *Biochemistry 11*: 4072-4084.

Goldstein, L., Pecht, M., Blumberg, S., Atlas, D., and Levin, Y. (1970). Water-insoluble enzymes. Synthesis of a new carrier and its utilization for preparation of insoluble derivatives of papain, trypsin, and subtilopeptidase A. *Biochemistry 9*: 2322-2334.

Guiseley, K. B., and Renn, D. W. (1977). *Agarose: Purification, Properties, and Biomedical Application*. Marine Colloids, Inc., Rockland, Maine.

Hjerten, S. (1964). The preparation of agarose spheres for chromatography of molecules and particles. *Biochim. Biophys. Acta 79*: 393-398.

Holmbergh, O. (1933). Über die Adsorption von α-Amylase aus Malz an Stärke. *Biochem. Z. 258*: 134-140.

Holmes, W. M., Hurt, R. E., Reid, B. R., Rimerman, R. A., and Hatfield, G. A. (1975). Separation of transfer ribonucleic acid by Sepharose chromatography using reverse salt gradients. *Proc. Nat. Acad. Sci. U.S. 72*: 1068-1071.

Horisberger, M. (1976). Immobilization of protein and polysaccharides on magnetic particles: Selective binding of microorganisms by concanavalin A-magnetite. *Biotechnol. Bioeng. 18*: 1647-1651.

Inman, J. K., and Dintzis, H. M. (1969). The derivatization of cross-linked polyacrylamide beads. Controlled introduction of functional groups for the preparation of special-purpose, biochemical adsorbents. *Biochemistry 8*: 4074-4082.

Johansson, J., and Lindberg, H. (1979). Chromatographic properties of Sephacryl S-300 superfine. *J. Biochem. Biophys. Methods 1*: 37-44.

Kennedy, J. F. (1974). Chemically reactive derivatives of polysaccharides. *Advan. Carbohydr. Chem. Biochem. 29*: 305-405.

Knight, C. S. (1967). Some fundamentals of ion-exchange-cellulose design and usage in biochemistry. *Advan. Chromatogr. 4*: 61-110.

Krämer, D. M., Lehmann, K., Plainer, H., Reisner, W., and Sprössler, B. G. (1974a). Enzymes covalently bound to acrylic gel beads. I. Interaction of hydrophilic anionic gel beads with biomacromolecules. *J. Polym. Sci., Polym. Symp. 47*: 77-87.

Krämer, D. M., Lehmann, K., Plainer, H., Reisner, W., and Sprössler, B. G. (1974b). Enzymes covalently bound to acrylic gel beads. II. Practical application of hydrolases covalently bound to hydrophilic anionic gel beads. *J. Polym. Sci., Polym. Symp. 47*: 89-94.

Krauss, G.-J., and Krauss, G. (1981). *Experimente zur Chromatographie*. VEB Deutscher Verlag der Wissenschaften, Berlin.

Kristiansen, T. (1974). Studies on blood group substances. V. Blood group substance a coupled to agarose as an immunosorbent. *Biochim. Biophys. Acta 362*: 567-574.

Kuga, S. (1980). New cellulose gel for chromatography. *J. Chromatogr. 195*: 221-230.

Låås, T. (1975). Agar derivatives for chromatography, electrophoresis and gel-bound enzymes. A benzylated dibromopropanol cross-linked Sepharose as an amphophilic gel for hydrophobic salting-out chromatography of enzymes with special emphasis on denaturing disks. *J. Chromatogr. 111*: 373-387.

Lederer, E., and Lederer, M. (1957). *Chromatography. A Review of Principles and Applications*. Elsevier, Amsterdam.
Lerman, L. S. (1953). A biochemical specific method for enzyme isolation. *Proc. Nat. Acad. Sci. U. S. 39*: 232-236.
Madden, J. K., and Thom, D. (1982). Properties and interactions of polysaccharides underlying their use as chromatographic supports. In *Affinity Chromatography and Related Techniques*, T. C. J. Gribnau, J. Visser and R. J. F. Nivard (eds.). Elsevier Scientific, Amsterdam, pp. 113-129.
Manecke, G. (1964). Über serologisch wirkende Proteinharze und Enzymharze. *Naturwissenschaften 51*: 25-34.
Mosbach, K., and Andersson, L. (1977). Magnetic ferrofluids for preparation of magnetic polymers and their application in affinity chromatography. *Nature 270*: 259-261.
Peška, J., Štamberg, J., Hradil, J., and Ilavský, M. (1976). Cellulose in bead form. Properties related to chromatographic uses. *J. Chromatogr. 125*: 455-469.
Porath, J. (1962). Cross-linked dextran as molecular sieves. *Advan. Protein Chem. 17*: 209-226.
Porath, J. (1981). Development of modern bioaffinity chromatography (a review). *J. Chromatogr. 218*: 241-259.
Porath, J., Axen, R., and Ernback, S. (1967). Chemical coupling of proteins to agarose. *Nature 215*: 1491-1492.
Porath, J., Janson, J. C., and Låås, T. (1971). Agar derivatives for chromatography, electrophoresis and gel-bound enzymes. I. Desulfated and reduced cross-linked agar and agarose in spherical bead form. *J. Chromatogr. 60*: 167-177.
Porath, J., Låås, T., and Janson, J. C. (1975). Agar derivatives for chromatography, electrophoresis and gel-bound enzymes. III. Rigid agarose gels cross-linked with divinyl sulfone (DVS). *J. Chromatogr. 103*: 49-62.
Rees, D. A. (1972). Shapely polysaccharides. *Biochem. J. 126*: 257-273.
Regnier, F. E., and Noel, R. (1976). Glycerolpropylsilane bonded phases in the steric exclusion chromatography of biological macromolecules. *J. Chromatogr. Sci. 14*: 316-320.
Schutyser, J., Buser, T., Van Olden, D., Tomas, H., Van Houdenhoven, F., and Van Dedem, G. (1982). Synthetic polymers applied to macroporous silica beads to form new carriers for industrial affinity chromatography. In *Affinity Chromatography and Related Techniques*, T. C. J. Gribnau, J. Visser, and R. J. F. Nivard (eds.). Elsevier Scientific, Amsterdam, pp. 143-153.
Štamberg, J., Peška, J., Dautzenberg, H., and Philipp, B. (1982). Bead cellulose. In *Affinity Chromatography and Related Techniques*, T. C. J. Gribnau, J. Visser, and R. J. F. Nivard (eds.). Elsevier Scientific, Amsterdam, pp. 131-141.

Starkenstein, E. (1910). Über Fermenteinwirkung und deren Beeinflussung durch Neutralsalze. *Biochem. Z.* *24*: 210-218.

Štrop, P., Mikes, F., and Chytilova, Z. (1978). Hydrophobic interaction chromatography of proteins and peptides on Spheron 300. *J. Chromatogr.* *156*: 239-254.

Švec, F., Hradil, J., Čoupek, J., and Kálal, J. (1975). Reactive polymers. I. Macroporous methacrylate copolymers containing epoxy groups. *Angew. Makromol. Chem.* *48*: 135-143.

Švec, F., Hrudková, H., Horák, D., and Kálal, J. (1977). Reactive polymers. VIII. Reaction of the epoxide groups of the copolymer glycidyl methacrylate-ethylendimethacrylate with aliphatic amino compounds. *Angew. Makromol. Chem.* *63*: 23-36.

Turková, J. (1976). Immobilization of enzymes on hydroxyalkyl methacrylate gels. *Methods Enzymol.* *44*: 66-83.

Turková, J., and Seifertová, A. (1978). Affinity chromatography of proteases on hydroxyalkyl methacrylate gels with covalently attached inhibitors. *J. Chromatogr.* *148*: 293-297.

Turková, J., Hubálková, O., Křiváková, M., and Čoupek, J. (1973). Affinity chromatography on hydroxyalkyl methacrylates. I. Preparation of immobilized chymotrypsin and its use in the isolation of proteolytic inhibitors. *Biochim. Biophys. Acta* 322: 1-10.

Turková, J., Blaha, K., Malaniková, M., Vancurová, D., Švec, F., and Kálal, J. (1978). Methacrylate gels with epoxide groups as supports for immobilization of enzymes in pH range 3-12. *Biochim. Biophys. Acta* *524*: 162-169.

Weetall, H. H. (1973). Affinity chromatography. *Sep. Purif. Methods* *3*: 199-229.

Weetall, H. H., and Filbert, A. M. (1974). Porous glass for affinity chromatography applications. *Methods Enzymol.* *34*: 59-72.

Weliky, N., and Weetall, H. H. (1965). The chemistry and use of cellulose derivatives for the study of biological systems. *Immunochemistry* *2*: 293-322.

Young, N. M., and Leon, M. A. (1978). Preparation of affinity chromatography media from soluble polysaccharides by cross-linkage with divinyl sulfone. *Carbohydr. Res.* *66*: 299-302.

Zeichner, M., and Stern, R. (1977). Resolution of ribonucleic acids by Sepharose 4B column chromatography. *Biochemistry* *16*: 1378-1382.

4
Chemical Modification of Supports and Immobilization of Affine Ligands

The covalent linkage of biologically active compounds on the support is normally a two step reaction consisting of:

1. Activation or functionalization of the chemically inert support
2. Coupling of the ligand to this modified matrix

Besides the proper selection of the support matrix, optimal covalent linkage of the ligand is essential. Frequently, the failure of affinity chromatographic purification is due to the wrong choice of conditions. In particular, the following factors must be considered:

Stability of the covalent linkage between support and ligand
Steric conditions in the surroundings of the ligands, for example, the distance of the affine groups of the ligand from the support surface
The type and site of covalent linkage; for complex formation, for example, essential groups must not be sterically or chemically blocked

To date, a variety of activation and immobilization methods has been tested and has, in part, found broad application. Without enough experience, however, it cannot be decided whether all these variants will become practically relevant. In this chapter several methods are discussed in more detail, methods developed for certain types of supports, as well as those used predominantly to introduce functional groups.

POLYSACCHARIDES

The polysaccharides have been widely used as chromatographic supports for many methods. The covalent linkage of the bioaffine ligand necessary for affinity chromatography is achieved largely via an activation or functionalization of the primary and/or secondary OH groups.

Cyanogen Bromide Activation

Since its introduction, cyanogen bromide activation (Axen et al., 1967) has been one of the most usual methods for CNBr activation and is to date the method of choice on the laboratory scale. The method has many advantages. The matrix material is universally applicable. Activation is achieved via free OH groups of the polysaccharide matrix. Besides the polysaccharides, synthetic polymers with accessible OH groups, such as Spheron or Separon, can also be activated (Turková et al., 1973; Turková, 1976).

The ligands are bound to the activated support by means of a primary aliphatic or aromatic amino group, which should be present in unprotonized form. The ligand may be a simple low-molecular-weight compound or a high-molecular-weight biopolymer with quaternary structure, such as hemoglobin (Chua and Bushuk, 1969).

The activation procedure can be carried out relatively simply. The degree of activation is reproducible and controllable; it is proportional to the amount of CNBr and the pH of activation. The physical structure of the matrix does not vary within broad limits of the degree of activation. Shrinking occurs only under extremely high activation as a result of cross-linking (Axen et al., 1967).

Extensive studies have been published on the mechanism of CNBr activation of polysaccharides (Axen and Ernback, 1971; Ahrgren et al., 1972; Bartling et al., 1972; Lowe and Dean, 1974; Wilchek et al., 1975; Kohn and Wilchek, 1978, 1982). Despite these studies, there has long been no definitive explanation of the exact mechanism, the nature of the active species, and the covalent bonding to the ligand. A cyclic imidocarbonate as well as a cyanate ester have been discussed as active species.

Understanding of the activation mechanism has been complicated by the fact that the covalent bonding to the various polysaccharides proceeds via different intermediate steps. The extensive studies by Kohn and Wilchek (1982) have led to further insight into the mechanism. They developed analytical procedures for quantitative determination of active species on CNBr-activated polysaccharides: imidocarbonates by selective acid hydrolysis followed by the determination of liberated ammonia and cyanate ester by a color reaction with pyridine and dimethylbarbituric acid.

Thus, activation proceeds according to the scheme shown in Figure 4.1. Cyanate ester and imidocarbonate are both active species. The predominant species on agarose is cyanate ester (70-85% of total coupling capacity), and that on cross-linked dextran and cellulose is a cyclic imidocarbonate in position 2,3 or 3,4 of the carbohydrate ring. This different behavior can be explained by steric factors. Agarose does not contain vicinal hydroxylic groups, and imidocarbonate can be produced only via 4,6 position or via interchain reaction. The stability of the two active species is a function of pH but with opposite dependence. In this way it was also possible to analyze the individual structures.

At low pH the imidocarbonate is extremely labile, but it is stable in the alkaline region, especially above pH 12. Under coupling conditions (pH 9; 4°C), more than 50% of the active groups still survive after 24 hr. The cyanate ester, however, is hydrolyzed at pH 14 and 20°C within a few seconds; in 0.1 N HCl, 50% of the initial quantity is still present after 24 hr.

The different structure and different pH dependence of the active species dominating in any given case explain the long-known discrepancy that activated agarose must be washed with 1 mM HCl and activated cross-linked dextran or cellulose, with 0.1 M $NaHCO_3$.

Cyanogen bromide-activated polysaccharides are not very stable in aqueous milieu. The active species hydrolyze in the inert carbamates and cyclic carbonate, respectively (Fig. 4.1). Therefore, coupling of the ligand should be made immediately after activation. On covalent binding of NH_2 groups containing ligands, as a rule, three different N-substituted derivatives may be formed: imidocarbonates, isourea and/or carbamates (Fig. 4.1). There is chemical evidence that N substituted isoureas are formed as the main product (Svensson, 1973; Wilchek et al., 1975). Several standard procedures are used for the activation and subsequent ligand binding.

The classic method is carried out in aqueous solution without a buffer at pH 10-11. The pH is maintained by the addition of sodium hydroxide solution and controlled by a pH meter. A simplified method was described by Porath and colleagues (1973). Since an alkaline phosphate buffer with very high buffer capacity (2-5 M) is used, control or adjustment of pH at activation is no longer required. A further simplification was developed by March and coworkers (1974). Again, pH control is not required because of the use of 2 M sodium carbonate. However, CNBr is used as a solution in acetonitrile. These solutions are stable at -20°C for long periods of time. When adding this solution to the support suspension, the CNBr is a finely dispersed precipitate and, therefore, can react very rapidly with the support.

N-Methyl-2-pyrrolidone can also be used as a suitable solvent for cyanogen bromide (Nishikawa and Bailon, 1975). These authors also found that the temperature of activation must be kept between 4 and

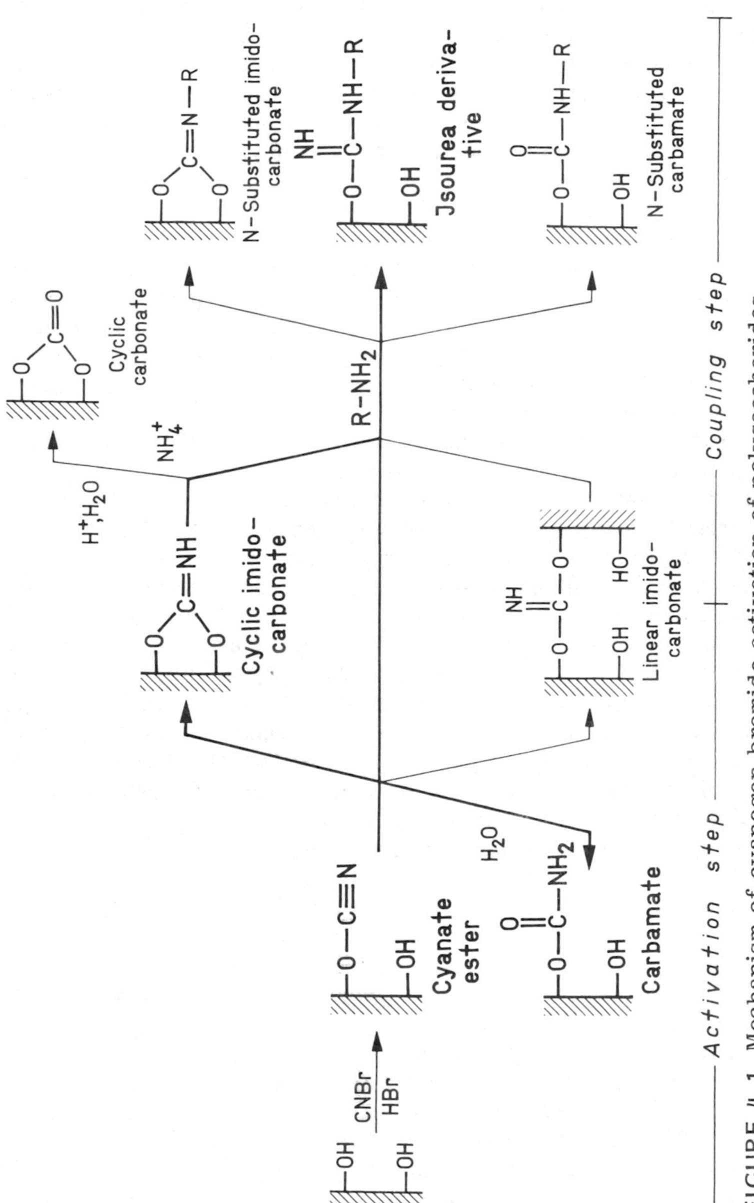

FIGURE 4.1 Mechanism of cyanogen bromide activation of polysaccharides.

10°C, but the coupling step that follows can be carried out at room temperature.

The optimal conditions for activation and subsequent ligand binding depend on many factors. Generally, high ligand concentrations and short reaction times give better results than low concentrations and long periods of incubation.

Few comments have been made about the influence of contaminated cyanogen bromide, although this could be of decisive importance. Cyanogen bromide must not contain free cyanide ions (Pommerening et al., 1979a). Several supports, such as poly-2-hydroxyethylmethacrylates (Spheron) or cellulose, but not Sepharose or Sephadex, adsorb cyanide ions that cannot be removed by even thorough washing. If these supports are incubated after activation with cyanogen bromide containing cyanide ions with substances or ions showing high affinity for cyanide ions, such as methemoglobin or Fe^{2+}, then as a result of complexing, desorption from the support takes place and can inhibit the fixed or incubated substances.

Furthermore, yellowish CNBr samples are poorly reactive. Lowe and Dean (1974) assume that a low concentration of trimers of CNBr, tribromotriazine, is present in the cyanogen bromide and is involved in the activation of the matrix. In aqueous solution, CNBr-activated polysaccharides are unstable. However, CNBr-activated Sepharose, a commercial product from Pharmacia (Uppsala, Sweden), after freeze-drying in the presence of dextran and lactose, can be stored longer than 18 months at a temperature below 8°C.

Other Methods

Triazine Method

The covalent linkage of ligands to OH groups containing supports using the triazine method is based on studies by Kay and coworkers (Kay and Crook, 1967; Kay and Lilly, 1970). The first step is activation of the support by cyanuric chloride (2,4,6-trichloro-s-triazine) or its derivatives, in which a chlorine is replaced by solubilizing groups (Fig. 4.2). In a second step the ligand is covalently bound via a primary NH_2 group by nucleophilic substitution of another chlorine. This method is of great importance for the binding of reactive dyes on polysaccharides, such as Cibacron Blue F3G-A (Chap. 12).

Periodate Oxidation

The oxidation of polysaccharides with sodium periodate for the attachment of proteins was first described by Sanderson and Wilson (1971). When studying the binding mechanisms, however, it should be considered that, analogously with CNBr, the polysaccharides react differently

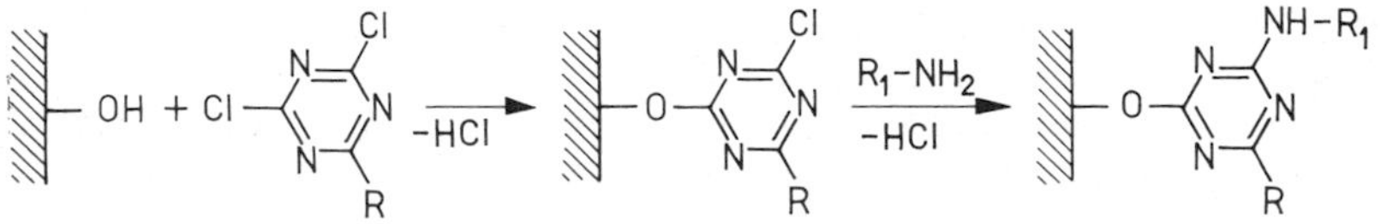

FIGURE 4.2 Covalent binding of ligands by the triazine method. R = -Cl, $-NH_2$, $-NH-CH_2-COOH$, $-O-CH_2-COOH$; R_1 = biomacromolecule, low-molecular-weight compound.

with the periodate, depending on the number of neighboring OH groups (Bouveng and Lindberg, 1960).

The very reactive aldehyde groups formed are present, in part, in hydrated inter- or intramolecular hemialdalic or hemiacetalic form. They can be coupled under very mild conditions with primary amino groups forming carbinol amines or the corresponding Schiff's bases (Fig. 4.3). The subsequent reduction with sodium borohydride stabilized the covalent bond and the support in general, as well as deactivated the noncoupled aldehyde groups. This is very important, because dialdehyde polysaccharides are unstable in alkaline solution (Guthrie, 1961).

This method is not suitable for less cross-linked dextran (Sephadex G75 to G200) because of the instability of the oxidized support, but this disadvantage can be used with great success for the solubilization of matrix-bound systems, thus considerably facilitating the investigation of these systems (Pommerening et al., 1973).

Due to the development of bead cellulose as a support this method becomes more and more important because, in contrast to the CNBr activation, the periodate-activated bead cellulose is stable at 4°C for at least several months (Turková et al., 1979; Pommerening, unpublished). Periodate oxidation can also be used for the covalent binding of ligands with carbohydrate residues, such as $NADP^+$ or other cofactors (Lamed et al., 1973), or for the binding of ribonucleic acids to amino (Gilman, 1971) or acylhydrazide residues on a modified agarose (Robberson and Davidson, 1972). The vicinal 2', 3'-diol groups of the ribose is readily oxidized by periodate. The method cannot be used for 2'-deoxyribonucleotides, which lack the diol systems.

Benzoquinone Method

In various cases benzoquinone has been used for the activation of OH groups in supports (Brandt et al., 1975). The binding of ligands takes place via primary amino groups at alkaline medium, as is shown in Figure 4.4. The first and second steps are the activation and the third

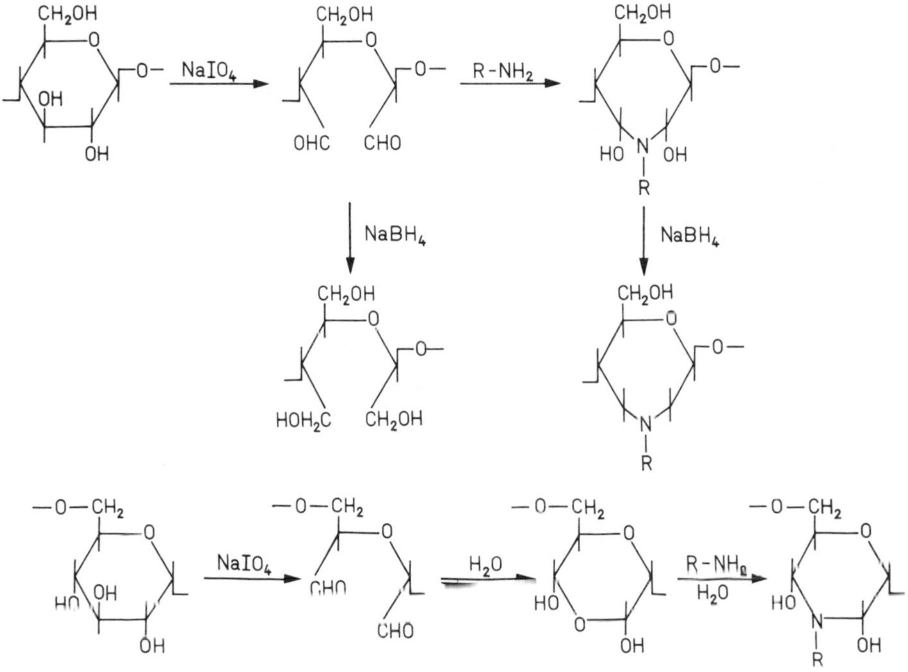

FIGURE 4.3 Covalent binding of ligands after periodate oxidation of polysaccharides.

step is the coupling procedure. The advantages of this method are that high amounts of fixed substances are obtained, the conjugates formed are very stable, and the coupling procedure can be carried out over a broad pH range.

Epoxides

The reaction of OH groups with epichlorohydrin (Fig. 4.5) leads to activated supports with oxiran groups (Porath and Fornstedt, 1970). For the introduction of the oxiran group, bisoxirans may also be used, such as 1,4-bis-(2,3-epoxypropoxy)butane (Sundberg and Porath, 1974) simultaneously with the introduction of a long hydrophilic spacer arm (Fig. 4.5). The simultaneous cross-linking increases stability but at the same time decreases flexibility and permeability.

Another possibility is the copolymerization of glycidylmethacrylate (Švec et al., 1975). The oxiran groups can react, in dependence on

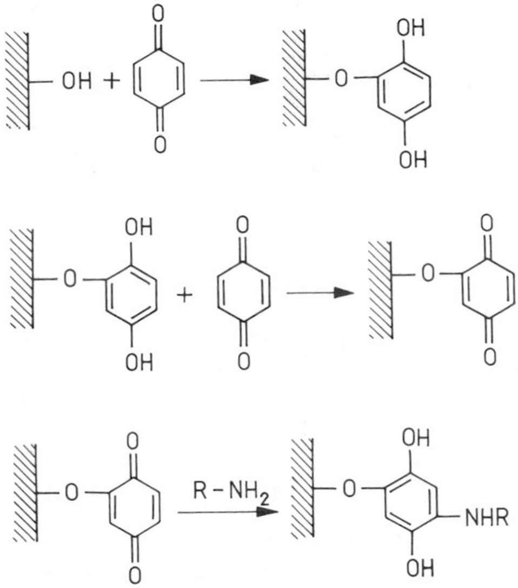

FIGURE 4.4 Covalent binding of ligands by means of the benzoquinone method.

pH, with different groups. In alkaline medium these are amino groups (Landt et al., 1978), hydroxyl groups (Vretblad, 1976), or thiol groups (Simons and Vander Jagt, 1977). In acidic medium the oxiran groups reacted with carboxylic groups (Turková et al., 1978). A disadvantage of the method is the relatively low activity of the oxiran ring. In most cases the conditions are pH 9-13, 16 hr reaction time, and temperature at 20-45°C. Therefore, this method is useful for the immobilization of small, stable ligands. Organic solvents may also be used as solvents for coupling solution (dimethylformamide or dioxane, both mixing with 50% water).

POLYACRYLAMIDES

Polyacrylamide gels (such as Bio-Gel from Bio-Rad, Richmond, California) can be converted into a support for affinity chromatography by derivatization of the carboxamide side chains (Inman and Dintzis, 1969; Inman, 1974). By reaction with a large excess of ethylenediamine at 90°C or hydrazine at 50°C, the amido group can be displaced by the aminoethyl or hydrazide group (Fig. 4.6). Thus these derivatives are the basis for the activation in different ways: introduction of aromatic groups, amines, and Curtius degradation. Another way

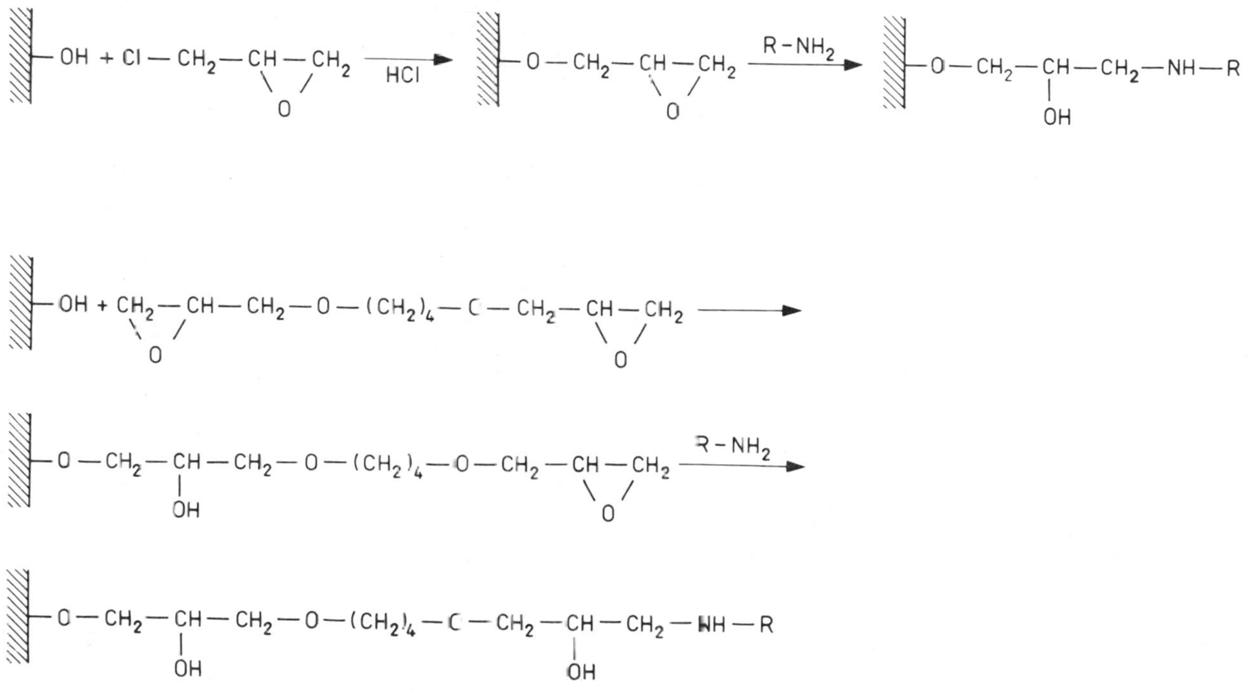

$$\text{—OH} + \text{Cl—CH}_2\text{—CH—CH}_2\ (\text{epoxide O}) \xrightarrow{\text{HCl}} \text{—O—CH}_2\text{—CH—CH}_2\ (\text{epoxide O}) \xrightarrow{\text{R-NH}_2} \text{—O—CH}_2\text{—CH(OH)—CH}_2\text{—NH—R}$$

$$\text{—OH} + \text{CH}_2\text{—CH—CH}_2\text{—O—(CH}_2)_4\text{—O—CH}_2\text{—CH—CH}_2\ (\text{bis-epoxide}) \longrightarrow$$

$$\text{—O—CH}_2\text{—CH(OH)—CH}_2\text{—O—(CH}_2)_4\text{—O—CH}_2\text{—CH—CH}_2\ (\text{epoxide O}) \xrightarrow{\text{R-NH}_2}$$

$$\text{—O—CH}_2\text{—CH(OH)—CH}_2\text{—O—(CH}_2)_4\text{—O—CH}_2\text{—CH(OH)—CH}_2\text{—NH—R}$$

FIGURE 4.5 Covalent binding of ligands by means of epichlorohydrin or bisoxiranes.

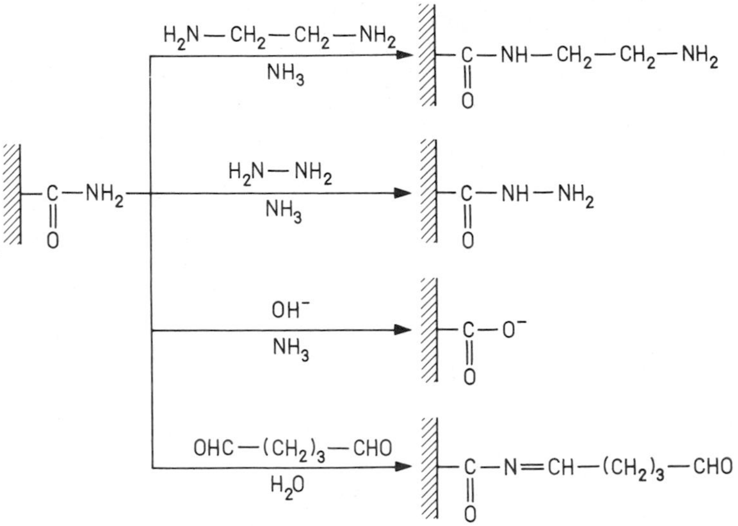

$$\text{—C(=O)—NH}_2 \xrightarrow[NH_3]{H_2N—CH_2—CH_2—NH_2} \text{—C(=O)—NH—CH}_2\text{—CH}_2\text{—NH}_2$$

$$\text{—C(=O)—NH}_2 \xrightarrow[NH_3]{H_2N—NH_2} \text{—C(=O)—NH—NH}_2$$

$$\text{—C(=O)—NH}_2 \xrightarrow[NH_3]{OH^-} \text{—C(=O)—O}^-$$

$$\text{—C(=O)—NH}_2 \xrightarrow[H_2O]{OHC—(CH_2)_3—CHO} \text{—C(=O)—N=CH—(CH}_2)_3\text{—CHO}$$

FIGURE 4.6 The first step of derivatization of polyacrylamides.

is the hydrolysis of the amide groups to yield carboxyl groups. A method for direct activation of polyacrylamide supports has been shown by Weston and Avrameas (1971). Glutaraldehyde can be reacted by one of its aldehyde groups with the amide group of the support, and the other aldehyde group is free for the binding of a ligand. The nature of the reaction between glutaraldehyde and polyacrylamide gels is not clear.

POROUS GLASS

The first step of the functionalization of porous silica glass is silanization (Weetall and Filbert, 1974). This is a chemical reaction between OH groups of the glass surface and an amino-functional trialkoxysilane, such as 3-aminopropyltriethoxysilane (Fig. 4.7). In further steps the amino group can be modified or activated for immobilization of a ligand by carbodiimide, thiophosgene, p-nitrobenzylchloride, glutaraldehyde, and others.

GENERAL REACTIONS FOR COUPLING LIGANDS

Many further activation methods have been developed for the modification or activation of a functional group by a polymer-analogous

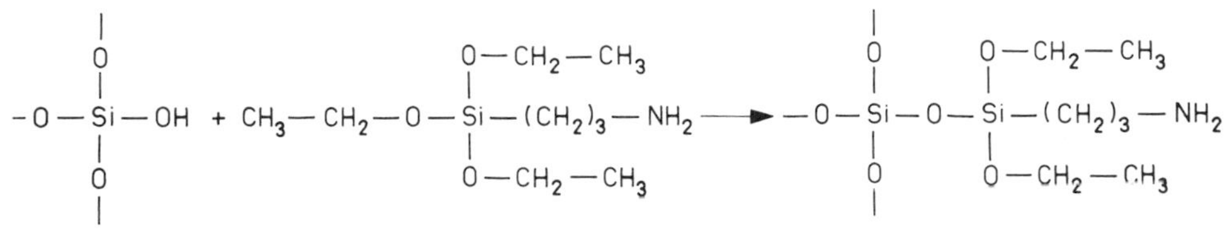

$$-O-Si(O-)_2-OH + CH_3-CH_2-O-Si(O-CH_2-CH_3)_2-(CH_2)_3-NH_2 \longrightarrow -O-Si(O-)_2-O-Si(O-CH_2-CH_3)_2-(CH_2)_3-NH_2$$

FIGURE 4.7 Silanization of glass and its derivatives.

reaction. These reactions are largely independent of the carrier matrix itself. In most cases amino or carboxyl groups are modified.

Carbodiimide Condensations

The carbodiimide method may be used for the covalent linkage between a free amino group and a free carboxylic group forming a peptide bond, and it does not matter whether the NH_2 or COOH group is at the support.

The mechanism of the reaction is given in Figure 4.8. In the first step carbodiimide reacts with carboxyl groups at pH 4-5 to give the highly reactive O-acylisourea (Hoare and Koshland, 1966). Then the activated carboxyl group condenses with an amino group to yield the

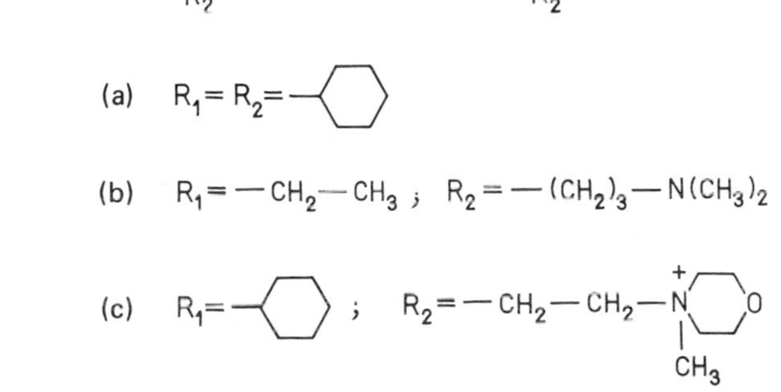

$$\text{—C(=O)—OH} + R_1-N{=}C{=}N-R_2 \xrightarrow[\text{pH 4-5}]{H^+} \text{—C(=O)—O—C(=N'H—}R_1\text{)—NH—}R_2 \xrightarrow{R_3-NH_2} \text{—C(=O)—NH—}R_3 + R_1-NH-C(=O)-NH-R_2$$

(a) $R_1 = R_2 = -C_6H_{11}$ (cyclohexyl)

(b) $R_1 = -CH_2-CH_3$; $R_2 = -(CH_2)_3-N(CH_3)_2$

(c) $R_1 = -C_6H_{11}$ (cyclohexyl) ; $R_2 = -CH_2-CH_2-N^+(CH_3)$ (morpholino, with O)

FIGURE 4.8 Covalent binding of ligands by carbodiimide coupling reaction.

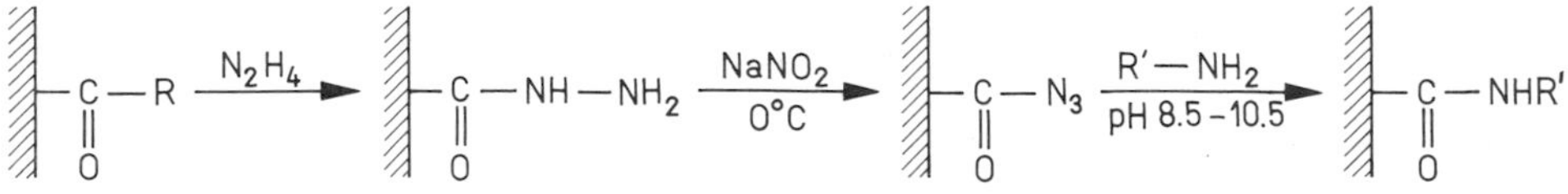

$$-\underset{\underset{O}{\|}}{C}-R \xrightarrow{N_2H_4} -\underset{\underset{O}{\|}}{C}-NH-NH_2 \xrightarrow[0°C]{NaNO_2} -\underset{\underset{O}{\|}}{C}-N_3 \xrightarrow[pH\ 8.5-10.5]{R'-NH_2} -\underset{\underset{O}{\|}}{C}-NHR'$$

FIGURE 4.9 Coupling of ligands by acylacides.

peptide bonding and the urea corresponding to the carbodiimide. The activated carboxyl group can also be reacted with OH or SH groups, forming carboxylate derivatives.

Several carbodiimides have been used successfully, such as the water-insoluble dicyclohexylcarbodiimide (Fig. 4.8a) and the water-soluble 1-ethyl-3-(3-dimethylaminopropyl)carbodiimide hydrochloride (EDC; Fig. 4.8b) or 1-cyclohexyl-3-(2-morpholinoethyl)carbodiimide metho-p-toluene sulfonate (CMC, Fig. 4.8c).

Acylation Reaction

The coupling of ligands with primary amino groups can be carried out via acylating groups on the polymers, such as acylazides or acid anhydrides. In the case of acylazide (Hornby et al., 1966), the carboxylic group is transferred into the corresponding hydrazide via the methyl ester or the carboxamide. Subsequently, the acylazide is formed with sodium nitrite via a Curtius rearrangement and then reacts with the primary amino groups (Fig. 4.9).

On the basis of a highly substituted carboxymethylcellulose, a number of supports for affinity chromatography has been prepared (Brümmer, 1974).

Coupling via the acid anhydride is not advantageous because free carboxylic groups arise at ligand binding simultaneously, thus also imparting to these supports properties of an ion exchanger.

Diazotation Procedures

Covalent linkage can also be achieved via aminoaryl side chains (Cuatrecasas, 1970; Inman and Dintzis, 1969; Weetall and Filbert, 1974). At first, the corresponding p-nitrophenyl group is introduced into the matrix and the NO_2 group converted by reduction into the NH_2-derivative (Fig. 4.10).

By treating the polymers with nitrous acid, the diazonium derivatives are prepared, which attack mainly the aromatic rings, such as phenol (tyrosine) or imidazole (histidine), to form the azo compounds. Historically, this is the oldest method of coupling proteins to diazotized polymers (Campbell et al., 1951). Although at present this

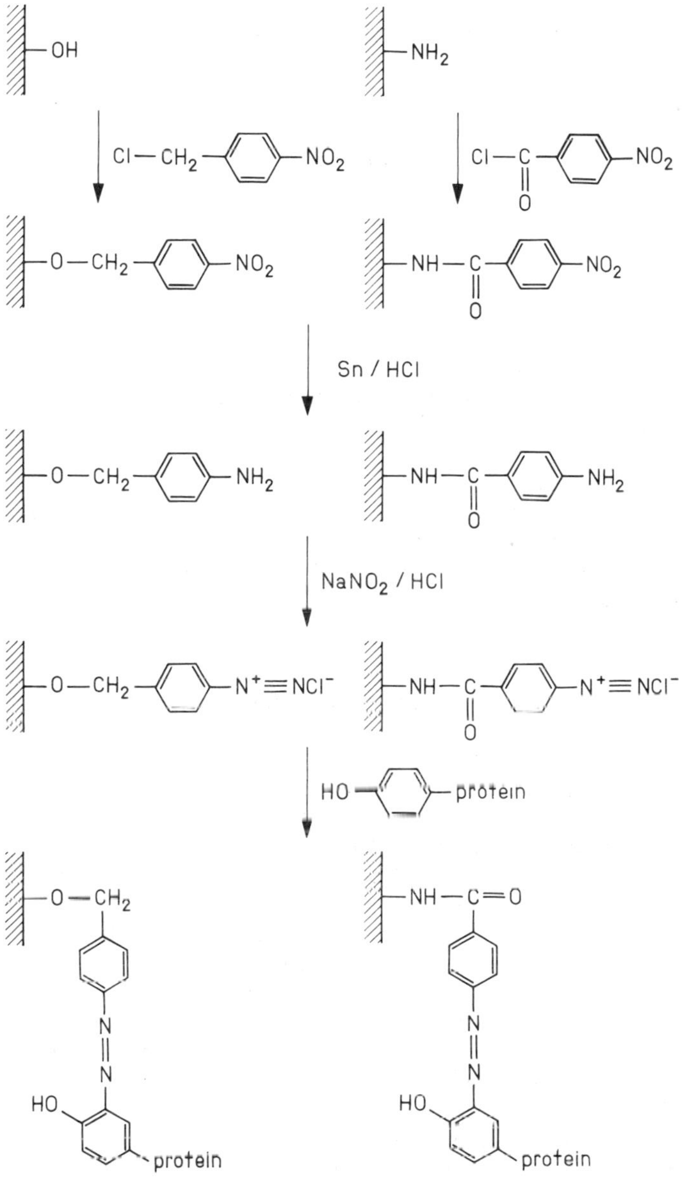

FIGURE 4.10 Covalent linkage of ligands to polymeric diazonium salts.

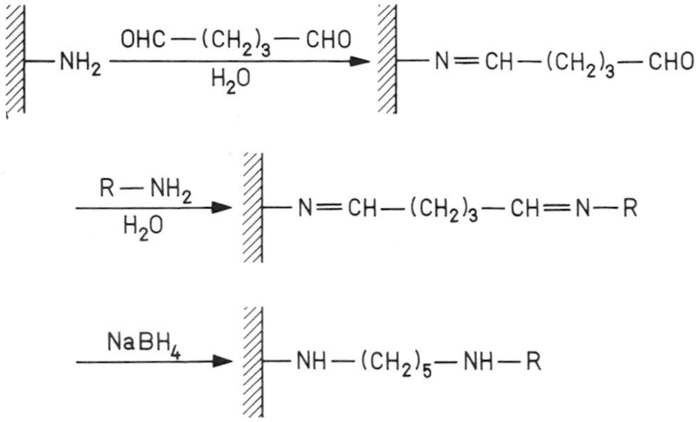

$$-NH_2 \xrightarrow[H_2O]{OHC-(CH_2)_3-CHO} -N{=}CH-(CH_2)_3-CHO$$

$$\xrightarrow[H_2O]{R-NH_2} -N{=}CH-(CH_2)_3-CH{=}N-R$$

$$\xrightarrow{NaBH_4} -NH-(CH_2)_5-NH-R$$

FIGURE 4.11 Coupling reaction with glutaraldehyde.

method is not frequently used, it offers two essential advantages: (a) the bound ligand can easily be split off again by complete reduction. (b) This reductive cleavage allows intact protein-inhibitor conjugates to be isolated under mild conditions (Lowe and Dean, 1974).

Glutaraldehyde

Glutaraldehyde was first used as an intermolecular cross-linking agent (Sabatini et al., 1963), but it can also be used with good results for covalent binding of ligands to NH_2 groups containing matrices, at the same time achieving prolongation of the side chains beside the ligand (Fig. 4.11). One aldehyde group reacts with the amino group of the support to form an azo-methin or Schiff's base. The second aldehyde group couples with an amino group of a ligand. Reduction with borohydride gives stable secondary amines.

Other bifunctional groups can be used (Lowe and Dean, 1974); however, it should be considered that side reactions may occur, such as cross-linking of the support, and permeability may decrease drastically. Frequently used bifunctional reagents include bisoxiranes (Porath and Sundberg, 1970), divinyl sulfones (Porath and Sundberg, 1972), or hexamethylene diisocyanate (Ozawa, 1967). Another possibility of activation of NH_2 groups containing matrices is the conversion of the amino group into the isocyanate or isothiocyanate groups by means of phosgene or thiophosgene (Inman, 1974; Weetall and Filbert, 1974).

Activation with Carbonylating Reagents

In the last few years, activation of supports (cross-linked agarose, beaded cross-linked dextran, cellulose, and glycophase-coated glass

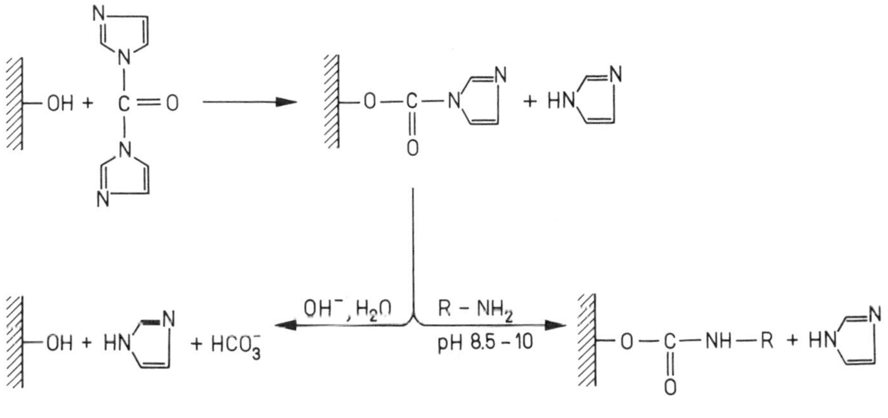

FIGURE 4.12 Reaction scheme of covalent fixation of ligands on polysaccharides by activation with 1,1'-carbonyldiimidazole.

beads) with carbonylating reagent, particularly 1,1'-carbonyldiimidazole (Bethell et al., 1979, 1981 a, b), has been described. The highly activated imidazolylcarbamate supports (Fig. 4.12) react with primary amino groups at pH 8.5-10.0 and are more stable to hydrolysis in comparison with N-hydroxysuccinimide ester-activated supports. The activated agarose is very stable when stored in dioxane. Under these conditions its half-life was longer than 14 weeks.

The N-alkylcarbamates that form by reaction of the amino group of a ligand to the support are more stable over a wide pH range compared with the isourea linkage obtained by cyanogen bromide activation. A further advantage of this method is that the N-alkylcarbamates obtained are uncharged in the pH range normally used. 1,1'-Carbonyldiimidazole is the most effective of the carbonylic reagents investigated. Activation with 1,1'-carbonyldi-1,2,4-triazole gives a much more reactive gel than with the imidazole derivatives; activation with 1,1'-carbonyldi-1,2,3-benzotriazole is ineffective.

Active Esters

Active esters can also be used with good success. Figure 4.13 shows some typical representatives. N-Hydroxysuccinimide esters of CH-Sepharose 4B can be used for spontaneous covalent coupling under mild conditions via primary amino groups even for very labile compounds (Cuatrecasas and Parikh, 1972). Besides the amino group, imidazole or SH groups can also react but to a lesser extent.

A p-nitrophenol ester of hydroxyalkylmethacrylate gels (NPAC) has been described by Turková (1976) and Čoupek and colleagues

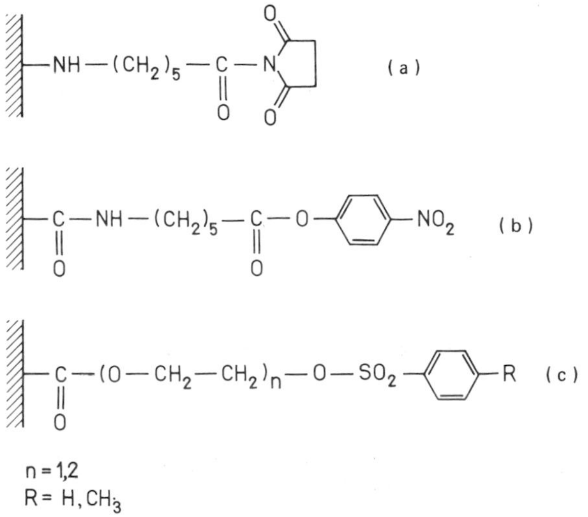

$-NH-(CH_2)_5-C(=O)-N$ (succinimide ring, two C=O) (a)

$-C(=O)-NH-(CH_2)_5-C(=O)-O-C_6H_4-NO_2$ (b)

$-C(=O)-(O-CH_2-CH_2)_n-O-SO_2-C_6H_4-R$ (c)

n = 1,2
R = H, CH_3

FIGURE 4.13 Structure of side chains with active ester groups: (a) N-hydroxysuccinimide, (b) p-nitrophenol, (c) arylsulfonates.

(1977). On the basis of reactive sulfoester groups, procedures for the binding of heterocyclic ligands, such as imidazole, have been developed (Kühn et al., 1981 a, b). These methods are well suited to the introduction of imidazole ligands via the annular nitrogen in position 1 and of spacer groups.

Succinylation

The reaction of succinic anhydride with matrices containing amino groups can be used to introduce a spacer arm and simultaneous free carboxylic groups instead of free amino groups (Cuatrecasas and Anfinsen, 1971). Groff and coworkers (1982) found, however, that matrices containing amino and hydroxyl groups show in addition to the expected N-succinylation a significant O-succinylation; for example, Sepharose CL-6B preactivated with (3-aminopropyl)triethoxysilane. A direct O-succinylation of polysaccharides is likewise possible and is a very simple way to introduce a spacer and carboxylic groups without prior activation. The ester bond is more labile than the amide bond in alkaline pH. The resulting succinylated Sepharose, for example, can be coupled with D-trytophan methyl ester — a competitive inhibitor of chymotrypsin — via the carbodiimide method and applied as a bioaffine matrix for chymotrypsin.

SPACERS

For a ligand with low molecular weight it is usually necessary for steric reasons that it not be bound directly to the carrier but via a side chain, generally called a spacer (Chap. 6). There are two principal possibilities for the synthesis of these carrier types:

1. The spacer is bound to the carrier and, subsequently, the bioaffine ligand to the end group of the spacer, using the various methods for matrix fixation already mentioned.
2. In the first step of synthesis the spacer is bound to the ligand as substituent according to the classic procedure of organic chemistry. The second step then includes matrix fixation via the side chain.

Both methods have advantages and disadvantages (O'Carra et al., 1974). The first method is used most frequently because it is simpler. Both steps of synthesis are performed on a solid phase. The disadvantage is the heterogeneity of the matrix synthesized. As a result of stepwise synthesis via a polymer-analogous reaction, there remain unsubstituted spacers at the matrix leading to unspecific bonds via hydrophobic or ionic interactions (Chap. 6). With a stepwise buildup of the spacer from different structural units, special attention must be paid to the problem of incomplete conversion.

For this reason, the second variant is preferred for several affinity matrices, for example, for matrices with cofactors as affine ligands (Brodelius et al., 1974). This synthetic approach guarantees a matrix support with an unequivocal ligand spacer system. The problems, however, are in the complicated procedures of purification and fractionation of the spacered ligands. The unspecific interaction does not depend only on the length and nature of the spacer, but also on the structure of the bonding sites. The formation of a positive amino nitrogen increased the possibility of disturbing the interaction.

A variety of spacers and their fixation to the matrix support has been described. Some characteristic examples are summarized in Figure 4.14.

Several supports already have a side chain, such as hydroxyalkylmethacrylate (Fig. 4.14a) or, generally, receive a spacer during functionalization, such as porous glass after the silanization process (Fig. 4.14b).

The dominating type of spacer used most widely is based on substituted alkyl side chains, with either ω,ω'-diaminoalkanes or ω-aminocarbonic acids bound to the support matrix (Fig. 4.14c, d). The most common are hexamethylendiamine and ε-aminocaproic acid. Corresponding types of supports are commercially available, such as

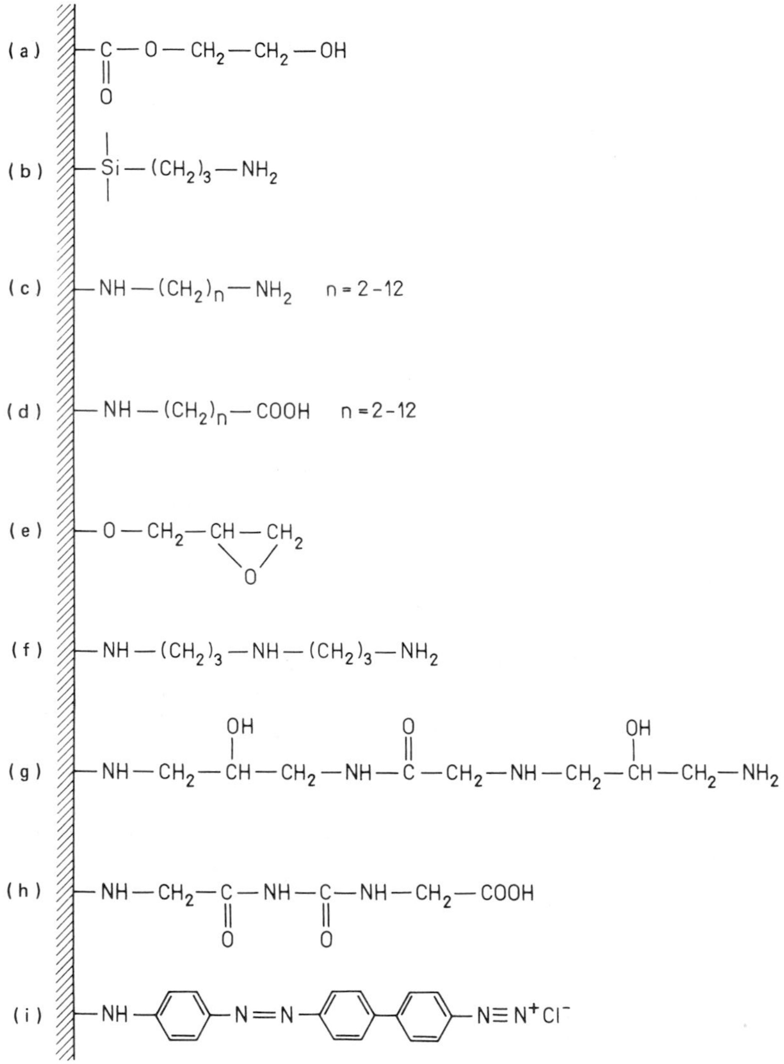

(a) $-C(=O)-O-CH_2-CH_2-OH$

(b) $-Si-(CH_2)_3-NH_2$

(c) $-NH-(CH_2)_n-NH_2$ $n = 2-12$

(d) $-NH-(CH_2)_n-COOH$ $n = 2-12$

(e) $-O-CH_2-CH-CH_2$ (epoxide: CH and CH_2 bridged by O)

(f) $-NH-(CH_2)_3-NH-(CH_2)_3-NH_2$

(g) $-NH-CH_2-CH(OH)-CH_2-NH-C(=O)-CH_2-NH-CH_2-CH(OH)-CH_2-NH_2$

(h) $-NH-CH_2-C(=O)-NH-C(=O)-NH-CH_2-COOH$

(i) $-NH-C_6H_4-N=N-C_6H_4-C_6H_4-N\equiv N^+Cl^-$

FIGURE 4.14 Some spacers used in affinity chromatography.

AH-Sepharose and CH-Sepharose. These spacers also show hydrophobic properties (Chaps. 6 and 15).

Another type are side chains obtained on activating the matrices with epichlorohydrin or bisoxirans (Fig. 4.14e). The OH groups increase the hydrophilic properties of the spacer. The OH group in β position to the ligand, however, may lead to steric complications during ligand binding (Pommerening et al., 1979b). Spacers with hydrophilic properties are based on the following structural units: 3,3'-diaminodipropylamine (Fig. 4.14f) (Harris et al., 1973), 1,3-diamino-2-propanol and bromoacetylation (Fig. 4.14g) (O'Carra et al., 1974), and oligopeptides (Fig. 4.14h) (Lowe et al., 1973). Even spacers with aromatic rings are used (Fig. 4.14i) (Lowe and Dean, 1971). The function of the spacer is discussed in detail in Chapter 6.

BLOCKING OF UNREACTED GROUPS

After covalent binding of the ligand it is necessary to block the unreacted active groups, that is, to deactivate the matrix. Otherwise, disturbances will occur during chromatography, such as covalent linkage of the substance to be purified. For methods that fix the ligands to the matrix via an amino group, especially suitable are low-molecular-weight primary amines. Thus, with cyanogen bromide activation, incubation with 2-aminoethanol (1 M) at pH 9 at room temperature for 2 hr is sufficient, but 4 hr is needed for epoxide-activated Sepharose. Other blodking agents are ethylamine, glycine, glucosamine, or 2-amino-2-hydroxymethylpropane-1,3-ol (Roche et al., 1975). The use of glycine, however, involves incorporation of charges that are undesirable in many cases.

In several cases, incubation with tris buffer, pH 8, for 1-2 hr is sufficient, as has been described for active esters. Unreacted NH_2 or COOH groups of spacers may be blocked using the carbodiimide method by different low-molecular-weight substances, such as 2-aminoethanol or acetic acid.

REFERENCES

Ahrgren, L., Kagedal, L., and Akerstrom, S. (1972). Covalent binding of proteins to polysaccharides by cyanogen bromide and organic cyanates. II. Model studies with methyl-4,6-benzylidene-α-D-glucopyranoside and cyanogen bromide. *Acta Chem. Scand. 26*: 285-288.

Axen, R., and Ernback, S. (1971). Chemical fixation of enzymes to cyanogen halide activated polysaccharide carriers. *Eur. J. Biochem. 18*: 351-360.

Axen, R., Porath, J., and Ernback, S. (1967). Chemical coupling of peptides and proteins to polysaccharides by means of cyanogen halides. *Nature 214*: 1302-1304.

Bartling, G. J., Brown, H. D., Forrester, L. J., Koss, M. T., Mather, A. N., and Stasiw, R. O. (1972). A study of the mechanism of cyanogen bromide activation of cellulose. *Biotechnol. Bioeng. 14*: 1039-1043.

Bethell, G. S., Ayers, J. S., Hancock, W. S., and Hearn, M. T. W. (1979). A novel method of activation of cross-linked agarose with 1,1'-carbonyldiimidazole which gives a matrix for affinity chromatography devoid of additional charged groups. *J. Biol. Chem. 254*: 2572-2574.

Bethell, G. S., Ayers, J. S., Hearn, M. T. W., and Hancock, W. S. (1981a). Investigation of the activation of cross-linked agarose with carbonylating reagents and the preparation of matrices for affinity chromatography purifications. *J. Chromatogr. 219*: 353-359.

Bethell, G. S., Ayers, J. S., Hearn, M. T. W., and Hancock, W. S. (1981b). Investigation of the activation of various insoluble polysaccharides with 1,1'-carbonyldiimidazole and of the properties of the activated matrices. *J. Chromatogr. 219*: 361-372.

Bouveng, H. O., and Lindberg, B. (1960). Methods in structural polysaccharide chemistry. *Advan. Carbohydr. Chem. 15*: 53-89.

Brandt, J., Andersson, L.-O., and Porath, J. (1975). Covalent attachment of proteins to polysaccharide carriers by means of benzoquinone. *Biochim. Biophys. Acta 386*: 196-202.

Brodelius, P., Larsson, P.-O., and Mosbach, K. (1974). The synthesis of three AMP-analogues: N^6-(6-aminohexyl)adenosine-5'-monophosphate, N^6-(6-aminohexyl)adenosine-2',5'-bisphosphate and N^6-(6-aminohexyl)adenosine-3',5'-bisphosphate and their application as general ligands in biospecific affinity chromatography. *Eur. J. Biochem. 47*: 81-89.

Brümmer, W. (1974). Affinitätschromatographie. *Kontakte (Merck)*: *1/74* 23-28, *2/74* 3-13.

Campbell, D. H., Luescher, E., and Lerman, L. S. (1951). Immunologic adsorbents. I. Isolation of antibody by means of a cellulose protein antigen. *Proc. Nat. Acad. Sci. U. S. 37*: 575-578.

Chua, G. K., and Bushuk, W. (1969). Purification of wheat proteases by affinity chromatography on hemoglobin-Sepharose column. *Biochem. Biophys. Res. Commun. 37*: 545-550.

Čoupek, J., Labský, J., Kálal, J., Turková, J., and Valentová, O. (1977). Reactive carriers of immobilized compounds. *Biochim. Biophys. Acta 481*: 289-296.

Cuatrecasas, P. (1970). Protein purification by affinity chromatography. Derivatizations of agarose and polyacrylamide beads. *J. Biol. Chem. 245*: 3059-3065.

Cuatrecasas, P., and Anfinsen, C. B. (1971). Affinity chromatography. *Methods Enzymol. 22*: 345-378.

Cuatrecasas, P., and Parikh, I. (1972). Adsorbents for affinity chromatography. Use of N-hydroxysuccinimide esters of agarose. *Biochemistry 11*: 2291-2299.

Gilman, P. T. (1971). The covalent binding of nucleotides, polynucleotides and nucleic acids to cellulose. *Methods Enzymol. 21*: 191-197.

Groff, J. L., Cherniak, R., and Jones, R. G. (1982). The incorporation of carboxyl groups into dextran and cross-linked agarose by O-succinylation. *Carbohydr. Res. 101*: 168-173.

Guthrie, R. D. (1961). The "Dialdehydes" from the periodate oxidation of carbohydrates. *Advan. Carbohydr. Chem. 16*: 105-158.

Harris, R. G., Rowe, J. J. M., Stewart, P. S., and Williams, D. C. (1973). Affinity chromatography of glucuronidase. *FEBS Lett. 29*: 189-192.

Hoare, D. G., and Koshland, D. E. (1966). A procedure for the selective modification of carboxyl groups in proteins. *J. Amer. Chem. Soc. 88*: 2057-2058.

Hornby, W. E., Lilly, M. D., and Crook, E. M. (1966). The preparation and properties of ficin chemically attached to carboxymethyl-cellulose. *Biochem. J. 98*: 420-425.

Inman, J. K. (1974). Covalent linkage of functional groups, ligands and proteins to polyacrylamide beads. *Methods Enzymol. 34*: 30-58.

Inman, J. K., and Dintzis, H. M. (1969). The derivatization of cross-linked polyacrylamide beads. Controlled introduction of functional groups for the preparation of special purpose, biochemical adsorbents. *Biochemistry 8*: 4074-4082.

Kay, G., and Crook, E. M. (1967). Coupling of enzymes to cellulose using chloro-s-triazines. *Nature 216*: 514-515.

Kay, G., and Lilly, M. D. (1970). The chemical attachment of chymotrypsin to water-insoluble polymers using 2-amino-4,6-dichloro s triazine. *Biochim. Biophys. Acta 198*: 276-285.

Kohn, J., and Wilchek, M. (1978). A colorimetric method for monitoring activation of Sepharose by cyanogen bromide. *Biochem. Biophys. Res. Commun. 84*: 7-14.

Kohn, J., and Wilchek, M. (1982). The determination of active species on CNBr and trichloro-s-triazine activated polysaccharides. In *Affinity Chromatography and Related Techniques*, T. C. J. Gribnau, J. Visser, and R. J. F. Nivard (eds.). Elsevier Scientific, Amsterdam, pp. 235-244.

Kühn, M., Pommerening, K., Beneš, M., Peška, J., and Štamberg, J. (1981a). Chemical transformations of polymers. XXI. Imidazole containing polymers from cross-linked macroporous poly-(methacrylic acid esters) with reactive sulfoester groups. *Angew. Makromol. Chem. 97*: 153-159.

Kühn, M., Pommerening, K., Mohr, P., Beneš, M., and Štamberg, J. (1981b). Chemical transformation of polymers. XXII. Poly-(methacrylic acid esters) with covalently linked imidazole side chains. *Angew. Makromol. Chem. 97*: 161-169.

Lamed, R., Levin, Y., and Wilchek, M. (1973). Covalent coupling of nucleotides to agarose for affinity chromatography. *Biochim. Biophys. Acta 231*: 231-235.

Landt, M., Boltz, S. C., and Butler, L. G. (1978). Alkaline phosphatase: Affinity chromatography and inhibition by phosphonic acids. *Biochemistry 17*: 915-919.

Lowe, C. R., and Dean, P. D. G. (1971). Affinity chromatography of enzymes on insolubilized cofactors. *FEBS Lett. 14*: 313-316.

Lowe, C. R., and Dean, P. D. G. (1974). *Affinity Chromatography*. John Wiley & Sons, London.

Lowe, C. R., Harvey, M. J., Craven, D. B., and Dean, P. D. G. (1973). Some parameters relevant to affinity chromatography on immobilized nucleotides. *Biochem. J. 133*: 499-506.

March, S. C., Parikh, I., and Cuatrecasas, P. (1974). A simplified method for cyanogen bromide activation of agarose for affinity chromatography. *Anal. Biochem. 60*: 149-152.

Nishikawa, A. H., and Bailon, P. (1975). Affinity purification methods. Improved procedures for cyanogen bromide reaction on agarose. *Anal. Biochem. 64*: 268-275.

O'Carra, P., Barry, S., and Griffin, T. (1974). Spacer arms in affinity chromatography: Use of hydrophilic arms to control or eliminate non-specific adsorption effects. *FEBS Lett. 43*: 169-175.

Ozawa, H. (1967). Bridging reagent for protein. I. The reaction of diisocyanates with lysine and enzyme proteins. *J. Biochem. (Tokyo) 62*: 419-423.

Pommerening, K., Blanck, J., Mauersberger, K., Behlke, J., Honeck, H., Smettan, G., Ristau, O., and Rein, H. (1973). Studies on the characterization of matrix-bound solubilized human hemoglobin. *Abh. Deut. Akad. Wiss. Berlin, Kl. Chem., Geol. Biol.* 179-186.

Pommerening, K., Jung, W., Kühn, M., and Mohr, P. (1979a). Problems of matrix activation in using cyanogen bromide methods. *J. Polym. Sci., Polym. Symp. 66*: 185-188.

Pommerening, K., Kühn, M., Jung, W., Buttgereit, K., Mohr, P., Štamberg, J., and Beneš, M. (1979b). Affinity chromatography of haemoproteins. 1. Synthesis of various imidazole containing matrices and their interaction with haemoglobin. *Int. J. Biol. Macromol. 1*: 79-88.

Porath, J., and Fornstedt, N. (1970). Group fractionation of plasma proteins on dipolar ion exchanger. *J. Chromatogr. 51*: 479-489.

Porath, J., and Sundberg, L. (1970). Biospecific adsorbents based on enzyme-inhibitor interaction. *Prot. Biol. Fluids 18*: 401-408.

Porath, J., and Sundberg, L. (1972). High capacity chemisorbents for protein immobilization. *Nature New Biol. 238*: 261-262.

Porath, J., Aspberg, K., Drevin, H., and Axen, R. (1973). Preparation of cyanogen bromide activated agarose gels. *J. Chromatogr. 86*: 53-56.

Robberson, D. L., and Davidson, N. (1972). Covalent coupling of ribonucleic acids to agarose. *Biochemistry 11*: 533-537.

Roche, A.C., Schauer, R., and Monsigny, M. (1975). Protein-sugar interactions. Purification by affinity chromatography of limulin, an N-acyl-neuraminidyl-binding protein. *FEBS Lett. 57*: 245-249.

Sabatini, D. D., Bensch, K., and Barrnett, R. J. (1963). Cytochemistry and electron microscopy. The preservation of cellular ultrastructure and enzymatic activity by aldehyde fixation. *J. Cell. Biol. 17*: 19-58.

Sanderson, C. J., and Wilson, D. V. (1971). Methods for coupling protein or polysaccharide to red cells by periodate oxidation. *Immunochemistry 8*: 163-168.

Simons, P. C., and Vander Jagt, D. C. (1977). Purification of glutathione-s-transferase from human liver by glutathione affinity chromatography. *Anal. Biochem. 82*: 334-341.

Sundberg, L., and Porath, J. (1974). Preparation of adsorbents for biospecific affinity chromatography. I. Attachment of group-containing ligands to insoluble polymers by means of bifunctional oxiranes. *J. Chromatogr. 90*: 87-98.

Švec, F., Hradil, J., Čoupek, J., and Kálal, J. (1975). Reactive polymers. I. Macroporous methacrylate copolymers containing epoxy groups. *Angew. Makromol. Chem. 48*: 135-142.

Svensson, B. (1973). Use of isoelectric focusing to characterize the bonds established during coupling of CNBr-activated amylodextrin to subtilisin type Novo. *FEBS Lett. 29*: 167-169.

Turková, J. (1976). Immobilization of enzymes on hydroxyalkyl methacrylate gels. *Methods Enzymol. 44*: 66-83

Turková, J., Bláha, K., Malinková, M., Vančurová, D., Švec, F., and Kálal, J. (1978). Methacrylate gels with epoxide groups as supports for immobilization of enzymes in pH range 3-12. *Biochim. Biophys. Acta 524*: 162-169.

Turková, J., Hubálková, O., Křiváková, M., and Čoupek, J. (1973). Affinity chromatography on hydroxyalkyl methacrylate gels. I. Preparation of immobilized chymotrypsin and its use in the isolation of proteolytic inhibitors. *Biochim. Biophys. Acta 322*: 1-9.

Turková, J., Vajčner, J., Vančurová, D., and Štamberg, J. (1979). Immobilization on cellulose in bead form after periodate oxidation and reductive alkylation. *Collect. Czech. Chem. Commun. 44*: 3411-3417.

Vretblad, P. (1976). Purification of lectins by biospecific affinity chromatography. *Biochim. Biophys. Acta 434*: 169-176.

Weetall, H. H., and Filbert, A. M. (1974). Porous glass for affinity chromatography application. *Methods Enzymol. 34*: 59-76.

Weston, P. D., and Avrameas, S. (1971). Proteins coupled to polyacrylamide beads using glutaraldehyde. *Biochem Biophys. Res. Commun. 45*: 1574-1580.

Wilchek, M., Oka, T., and Topper, Y. J. (1975). Structure of a soluble superactive insulin is revealed by the nature of the complex between cyanogen-bromide activated Sepharose and amines. *Proc. Nat. Acad. Sci. U.S. 72*: 1055-1058.

5
Choice of Affine Ligand

GENERAL ASPECTS

To be suitable as a ligand in affinity chromatography, a compound must fulfill two conditions:

1. It should rather specifically form a reversible complex with the substance to be purified.
2. The compound should possess a chemically modifiable group through which the covalent linkage to the matrix can occur.

Many compounds meet these two requirements. The wide limits of possible variation of the affinity chromatographic principle in its practical implementation afford the effectivity of the method. This effectivity results from

The variety of usable ligands that may be individually adapted to the given purpose
The use of group-specific ligands
The variability of the ligands that may be used for purification of a substance

A survey of some substance classes is given in Table 5.1. Affinity ligands represent rather different types of chemical compounds. Simple low-molecular-weight compounds can be used, such as substrates, as well as high-molecular-weight compounds, such as proteins with a quarternary structure and antibodies. For example, for the purification of proteases, amino acids and peptides, such as D-leucine for

TABLE 5.1 Some Classes of Affine Ligand for the Affinity Chromatography

Field of application	Ligand
Enzymology (enzymes, inhibitors, cofactors)	Enzymes, substrates, and their analogs, products, inhibitors, allosteric effectors, cofactors, antibodies, substances without biological function
Peptides (affinity labeled)	Enzymes, antibodies
Immunology	Haptens, antigens, antibodies
Protein chemistry (receptors, binding and transport proteins, glycoproteins)	Hormones, vitamins, steroids, lectins, lipids, antibodies, mono- and polysaccharides
Nucleic acids, nucleotides	Nucleic acid bases, nucleosides, oligonucleosides, dihydroxyboryl derivatives
Isolation and fractionation of cells and viruses	Antigens, antibodies, lectins, hormones

aminopeptidase (Turková et al., 1976) or L-tyrosyl-D-tryptophan for carboxypeptidase (Cuatrecasas et al., 1968), as well as biomacromolecules, such as hemoglobin for proteases from malted wheat flour (Chua and Bushuk, 1969) can be used.

It is also possible to use simple molecules without direct biological function. Matrix-bound imidazole can be used for the selective removal of hemoglobin (Pommerening et al., 1979), and reactive dyes have found broad application as group-specific ligands (Chap. 12).

The affinity of ligands can be varied over a broad range of force and specificity. Highly specific interactions are used, such as antibody antigen interaction, and interaction with lower affinity, for example, enzymes and substrates.

A ligand with a very narrow specificity – or monospecificity – has a high effectivity for the purification of a particular substance, but only for that substance, not for any other compound. Narrow specificity also has disadvantages because a special ligand-matrix combination is required for every substance to be purified. Therefore, ligands with group specificity, so-called general ligands, are of great practical value (Mosbach, 1974). They allow a variety of substances to be purified with a single affine matrix. In Table 5.2 some group-specific ligands are listed that are in part commercially available in immobilized form from different companies. These ready-to-use matrices have always contributed to the rapid development and broad application of affinity chromatography in laboratory practice.

Group specificity refers not only to functional properties but also to interaction with the same or analogous structural parameter (hydrophobic areas or sterically preformed binding sites).

In addition to their use in the purification of substances, the group-specific ligands are important for the elucidation of the molecular mechanism of interaction of the two partners. For example, if it is possible to immobilize the ligands via different chemical groups of the ligand, one can identify those structures that are essential for the binding by means of their adsorption behavior (Chap. 6). Biomacromolecules can specifically interact with different compounds or different structural sites. Therefore, it is possible in the purification of one compound to use different ligands adapted to the conditions given. For a given compound a more or less broad spectrum of affinity chromatographic procedures is known. For example, in Table 5.3, some possibilities for interferon purification by the affinity technique are summarized. Interferons are a heterogeneous group of molecules with antiviral activities. The best method is the use of immobilized antibodies. Other ligands with high selectivity utilize structural properties, such as hydrophobic areas, polynucleotide binding sites, carbohydrate moiety, and complex formation of transition metals with histidine and cysteine residues. The background of interaction on the molecular level is not in all cases clear. In the following sections several group-specific ligands and several special variants of application are discussed.

COENZYMES

The catalytic activity of many enzymes depends on the presence of coenzymes or cofactors. More than 600 (almost one-third) of the more than 2000 enzymes known so far require a nucleotide coenzyme. The reaction of both partners follows an ordered mechanism involving a stoichiometric enzyme-coenzyme complex. Therefore, the coenzymes, particularly the nucleotide cofactors, are very suitable group-specific ligands for dehydrogenases, kinases, and transaminases. In Figure 5.1 the chemical structures of the most important representatives of the adenine nucleotides and nicotine adenine dinucleotides are given. Especially, by the pioneering work of the groups around Mosbach, Dean, and Lowe, intensive investigations were carried out on the synthesis and applications of immobilized and nucleotide cofactors as group-specific ligands in the early 1970s. Covalent fixation can occur at different sites of the molecule:

- Via the amino group of the purine ring in position 6 (Lowe and Dean, 1971)
- Via the purine ring in position 8 by diazotation (Lowe and Dean, 1971)

TABLE 5.2 Summary of Some Group-Specific Ligands

Ligand	Specificity and applications
Coenzymes	
Pyridine and adenine nucleotides and their structural analogs	
Nicotinamide adenine dinucleotide (NAD^+)	Dehydrogenases
Nicotinamide adenine dinucleotide phosphate ($NADP^+$)	Dehydrogenases
Adenosine-5'-triphosphate (ATP)	Kinases
Adenosine-5'-monophosphate (5'-AMP)	NAD^+- and ATP-dependent enzymes (dehydrogenases, kinases)
Adenosine-2',5'-bisphosphate (2',5'-ADP)	$NADP^+$-dependent enzymes (dehydrogenases)
Flavin nucleotides	
Flavin mononucleotide (FMN), flavin adenine dinucleotide (FAD)	FMN- and FAD-dependent enzymes (flavokinase, reductases, oxidases)
Lectins	
Concanavalin A (Con A)	α-D-mannosyl, α-D-glucosyl, and sterically similar residues; isolation and purification of polysaccharides, glycoproteins, glycolipids, membrane vesicles, enzyme-antibody conjugates

Lentil lectin	α-D-glucosyl, α-D-mannosyl residues; isolation and purification of glycoproteins, cell surface antigens
Wheat germ lectin	N-acetyl-β-D-glucosaminyl residues; polysaccharides, glycoproteins, cells, subcellular particles
Helix pomatia lectin	N-acetyl-α-D-galactosaminyl residues; glycoproteins, cells (lymphocytes)
Sugars	
Monosaccharides and their derivatives	Lectins and sugar-binding proteins and enzymes
Nucleic acids	
Polyuridylic acid [poly(U)]	Poly(A), oligo(A) sequences; nucleic acids, mRNA, poly(U)-binding proteins (interferon)
Polyadenylic acid [poly(A)]	Poly(U), oligo(U) sequences; nucleic acids and oligonucleotides, RNA-binding proteins, antibodies
Oligodeoxythymidylic acid (oligo dT)	RNA, polyadenylic polymers
Other ligands	
3-Aminophenylboronic acid	Complex formation with unsubstituted coplanar cis diols; sugars and their derivatives, nucleosides, glycoproteins, serine proteases, ribonucleosides, coenzymes
Histone	DNA and histone binding proteins

TABLE 5.2 (continued)

Ligand	Specificity and applications
[Other ligands]	
Protein A	Complex formation with Fc-region of IgG and related molecules; IgG (mammalian species), IgG subclasses and fragments, immune complex, antigens
Lysine	Amphoteric character; ribosomal RNA, plasminogen, plasminogen activator, double-stranded DNA
Arginine	Amphoteric character (biospecific or charge-dependent affinity for L-arginine); fractionation of plasma proteins (prothrombin, fibronectin, prekallikrein)
Heparin	Polyanionic character, usually wide range of proteins; plasma proteins (coagulation proteins), lipases, lipoproteins, DNA and RNA polymerases, steroid receptors, protein synthesis factors, restriction endonucleases
p-Aminobenzamidine	Inhibitor or substrates
Ovomucoid	Isolation of proteases or their removal from other proteins
Hemoglobin	
Trypsin inhibitor (soybean)	
Imidodiacetic acid	Metal chelate; broad range of proteins and nucleotides
Acriflavin	Charge transfer complex; oligonucleotides, nucleotides, vitamins, plasmids

Reactive dyes	
Cibacron Blue F3G-A	Multivalent interaction; broad range of enzymes and proteins: NAD^+ ($NADP^+$)-requiring enzymes (dehydrogenases, kinases), plasma proteins (albumin, lipoproteins)
Procion Red HE-3B	Multivalent interaction; broad range of enzymes and proteins: $NADP^+$ (NAD^+)-requiring enzymes (dehydrogenases, reductases)
Reversible covalent chromatography	
Organomercurial derivatives	
p-Chloromercuribenzoate	SH groups; thiol-containing peptides, proteins, enzymes, mercurated polynucleotides
p-Aminophenylmercuric	
2-Pyridyldisulfides	
Glutathione	
2-Hydroxypropyl	
Thiol derivatives	
N-Acetyl-D,L-homocysteinsulfhydryl	
Cysteamine	
Hexanethiol	
Hydrophobic side chains	
n-Alkyl (octyl, hexyl)	Hydrophobic interaction
Phenyl	Broad range of biomacromolecules and other hydrophobic compounds

TABLE 5.3 Some Possibilities of Interferon Purification by Affinity Technique

Ligand	Reference
Antibody (anti-interferon)	Berg et al. (1978)
Monoclonal antibodies	Secher and Burke (1980)
Cibacron Blue F3G-A	Jankowski et al. (1976)
Con A	Davey et al. (1976)
Poly(U)	De Maeyer-Guignard et al. (1978)
Phenyl	Mizrahi et al. (1978)
Metal chelate	Edy et al. (1977)
Albumin	Huang et al. (1974)

Via the ribose residue after periodate oxidation (Lamed et al., 1973)
Via the terminal phosphate group (Harvey et al., 1974b)

All methods of covalent linkage are hindered by the relative instability and reactivity (Dean and Lowe, 1972). Furthermore, the function of immobilized cofactors as bioaffine ligands depends on the spacer length and on the position of covalent binding (Chap. 6). It was found that immobilization of the whole cofactor molecules is not a requirement. Frequently, fragments, structural analogs, or substances possessing partial structures, so-called half-molecules, are group-specific ligands as well.

The best ligands are adenosine-5'-monophosphate (5'-AMP) and adenosine-2',5'-bisphosphate (2',5'-ADP). Many working groups have intensively investigated these ligands (Lowe et al., 1972; Brodelius et al., 1974; Mosbach, 1974; Harvey et al., 1974a; Harvey et al., 1974b; Mosbach, 1978 a, b; Lowe, 1978). To obtain an effective matrix with them, it is first necessary to fix the structural analogs via the 6-carbon-long spacer to the matrix, and second, it is advantageous to avoid unspecific interactions due to the underivatized spacer arms, first to synthesize the spacer-ligand components and then to immobilize them.

In Figure 5.2, the partial structures of the two most common immobilized AMP analogs are given.

These ligands, with Sepharose 4B as matrix, are ready-to-use matrices from Pharmacia (Uppsala, Sweden) with the trade names

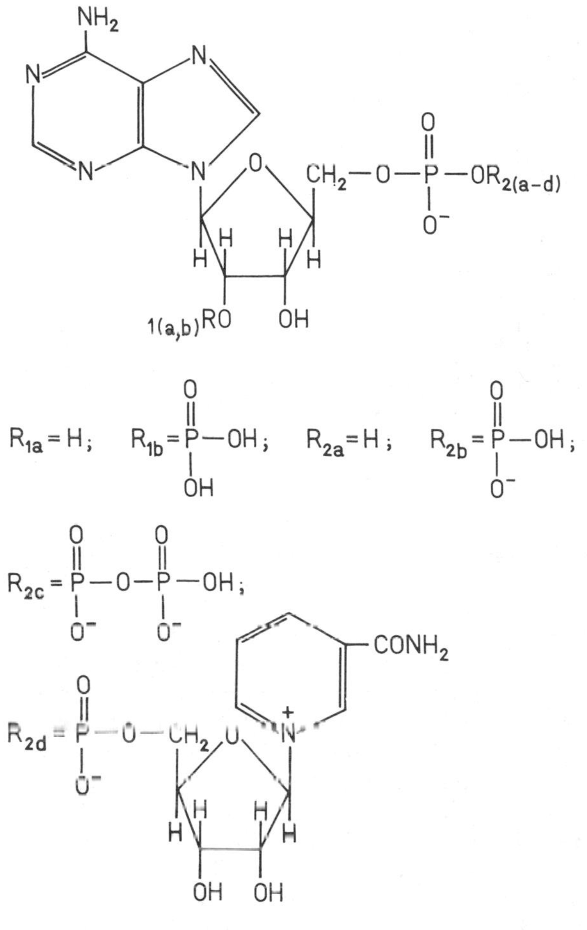

FIGURE 5.1 Chemical structures of adenine nucleotides and nicotinamide adenine dinucleotides. Adenosine-5'-monophosphate (AMP) = R_{1a}, R_{2a}; adenosine-5'-diphosphate (ADP) = R_{1a}, R_{2b}; adenosine-5'-triphosphate (ATP) = R_{1a}, R_{2c}; nicotinamide adenine dinucleotide (NAD^+) = R_{1a}, R_{2d}; nicotinamide adenine dinucleotide phosphate ($NADP^+$) = R_{1b}, R_{2d}.

5'-AMP-Sepharose 4B and 2',5'-ADP Sepharose 4B. The concentration of coupled ligands is approximately 2 µmol/ml swollen gel.

The efficiency of these matrix types was demonstrated quite early. Thus, Brodelius and Mosbach (1973) succeeded in separating five

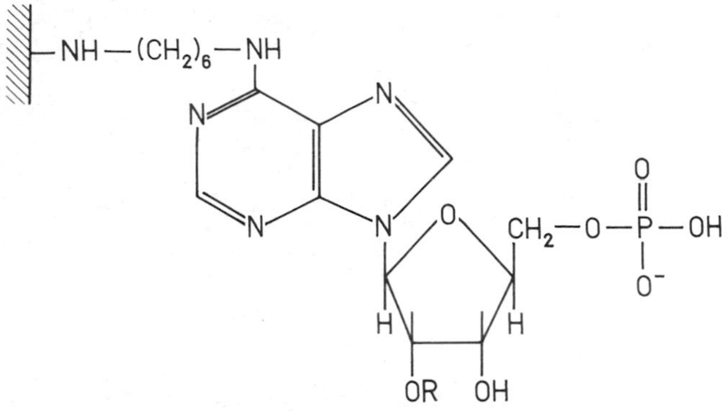

FIGURE 5.2 Partial structure of the two most common immobilized AMP analogs; R = H, 5'-AMP; R = PO_3H_2, 2',5'-ADP.

isoenzymes of lactate dehydrogenases by adsorption on immobilized AMP and subsequent desorption using a concave gradient of NADP (Fig. 5.3).

In model studies, Brodelius and colleagues (1974) demonstrated the selective adsorption and stepwise affinity elution of a mixture of lactate dehydrogenase, glucose-6-phosphate dehydrogenase, and phosphogluconate dehydrogenase on Sepharose-bound N^6-(6-aminohexyl)-adenosine-5'-monophosphate (Fig. 5.4a) and Sepharose-bound N^6-(6-aminohexyl)-2',5'-bisphosphate (Fig. 5.4b).

Only the $NADP^+$-dependent enzyme, 6-phosphogluconate dehydrogenase, was not adsorbed on the first matrix. The adsorbed NAD^+-dependent enzymes could be eluted with stepwise affinity elution. Figure 5.4b shows the behavior of the enzyme toward the other matrix. In this case the lactate dehydrogenase was not bound. The other enzymes could be eluted with 0.1 mM and 5 mM $NADP^+$ pulses, respectively.

The use of these types of matrix in the isolation and purification of membrane-bound detergent-solubilized enzymes has great advantages. Yasukochi and Masters (1976) described the purification of SDS-electrophoretically homogenous NADPH-cytochrome c (cytochrome P_{450}) reductase from detergent-solubilized rat and pig liver microsomes using 2',5'-ADP-Sepharose 4B. The degree of purification against the enzyme purified partially by DEAE-cellulose is 11.5. Isolation is also possible from the 105,000 *g* supernatant of solubilized pig liver microsomes (Fig. 5.5), but in this way only 40% activity was adsorbed. The initial broad peak also has reductase activity. This example shows that it is not always advantageous to carry out purification as a one-step affinity chromatographic procedure. Preliminary

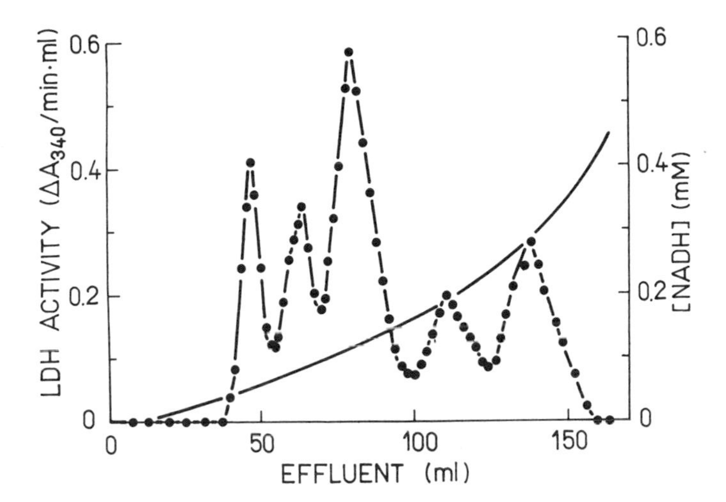

FIGURE 5.3 Elution of lactate dehydrogenase isoenzymes using a concave gradient of NADH. Protein (0.2 mg) in 0.2 ml of 0.1 M sodium phosphate buffer, pH 7.0, 1 mM β-mercaptoethanol, and 1 M NaCl was applied to an AMP-Sepharose column (140 X 6 mm, containing 2.5 g of wet gel), equilibrated with 0.1 M sodium phosphate buffer, pH 7.5. After washing the column with this buffer (10 ml), the isoenzymes were eluted with a concave gradient of 0.0-0.5 mM NADH in the same buffer, containing 1 mM β-mercaptoethanol and 1.0 ml fractions were collected at a rate of 3.4 ml/hr. (From Brodelius and Mosbach, 1973.)

purification by a conventional method is often both reasonable and necessary.

Dignam and Strobel (1977) have purified the same enzyme from rat liver by affinity chromatography on Sepharose containing $NADP^+$ as ligand via an adipic acid dihydrazide spacer arm. Linkage between the spacer and $NADP^+$ is achieved here via the ribose residue after periodate oxidation. The degree of purification with this matrix was also very good. The enzyme could be purified up to a specific activity of 69.4 (μmoles of cytochrome c reduced per minute per milligram of protein at 30°C in 0.3 M potassium phosphate buffer, pH 7.7).

LECTINS

Today, lectins are a very interesting and important group of affine ligands with a broad selective group specificity. They are a class of

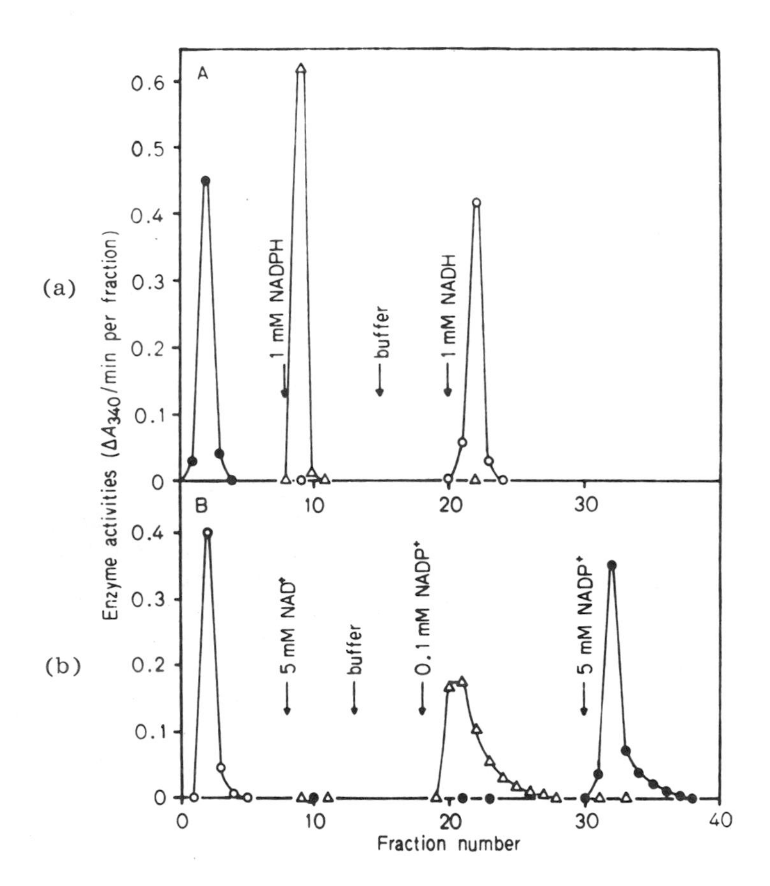

FIGURE 5.4 Affinity chromatography of mixtures of lactate dehydrogenase, glucose-6-phosphate dehydrogenase, and phosphogluconate dehydrogenase on Sepharose-bound N^6-(6-aminohexyl)adenosine-5'-monophosphate (a) and on Sepharose-bound N^6-(6-aminohexyl)adenosine-2',5'-bisphosphate (b). Applications of NAD(H) and NADP(H) dissolved in irrigant buffer are indicated by arrows: (o—o) Lactate dehydrogenase; (Δ—Δ) glucose-6-phosphate hydrogenase; (●—●) 6-phosphogluconate dehydrogenase. Bed volume: 0.3 ml; sample: approximately 0.1 U of each enzyme; buffer: 0.1 M tris-HCl, pH 7.6; fraction volume, 1.0 ml. (From Brodelius et al., 1974.)

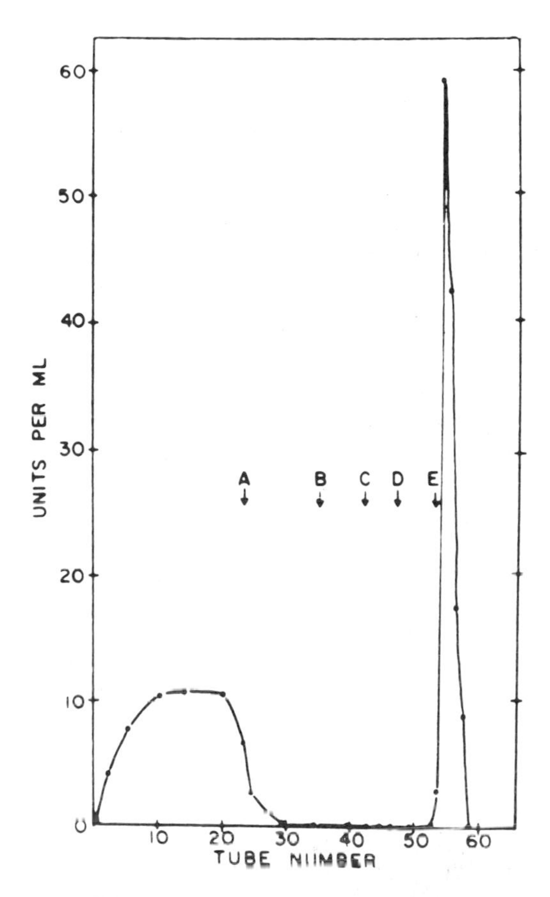

FIGURE 5.5 Affinity chromatography of detergent-solubilized pig liver microsomes on Sepharose 4B-bound $\underline{N}^6$-(6-aminohexyl)adenosine-2',5'-bisphosphate. The supernatant from a 105,000 $\underline{g}$, 60 min centrifugation of pig liver microsomes, solubilized with 1.0% Renex 690 and 0.5% sodium deoxycholate, and containing a total activity of 39.8 IU of NADPH-cytochrome $\underline{c}$ reductase in 50 ml was applied to the affine gel (1.0 X 2.5 cm), equilibrated with 10 mM potassium phosphate buffer, pH 7.7, containing 20% glycerol, 0.1% Renex 690, 0.02 mM EDTA, and 1 mM dithiothreitol. Applications of elution buffer are indicated by the arrows: (A) 100 mM potassium phosphate buffer, pH 7.7, containing 20% glycerol, 0.1% Renex 690, 0.2 mM EDTA, and 1 mM dithiothreitol; (B) equilibrating buffer; (C) 1 mM NADH in the same buffer as B; (D) equilibrating buffer; (E) 1 mM NADPH in the equilibrating buffer. (From Yasukochi and Masters, 1976.)

proteins from plants, invertebrates, and bacteria and form a complex with sugar residues of glucosides, oligo- and polysaccharides, glycoproteins, and glycolipids, and others. Complex formation is analogous to the interaction between antibody and antigen. There are great differences in the specificity of lectins, although most lectins do not have narrow specificity for one sugar only (Table 5.2). The most common lectin is concanavalin A (Con A); it reacts with α-D-mannopyranosyl, α-D-glucopyranosyl, and other sterically related residues. The hydroxyl groups in positions C_3, C_4, and C_6 are important for binding. Hydrogen bonding and electrostatic interactions are essentially involved in complex formation. Furthermore, the metal ions Ca^{2+} and Mn^{2+} are essential constituents of the sugar binding sites and are also essential for the tertiary structure.

Aspberg and Porath (1970) and Lloyd (1970) were the first to report application of immobilized lectins for the specific adsorption of glycoproteins. Today, affinity chromatography with immobilized lectins is an important step for the purification and isolation of polysaccharides, glycoproteins, glycolipids, cell fragments, and cells. In many cases it is this method that opens the possibility to prepare substances of sufficient purity and in sufficient quantity. The specific interaction with sugar residues is retained in the presence of detergent, such as 1% sodium deoxycholate. Therefore, lectins are very suitable ligands for the purification of membrane proteins, for example, for Con A receptor from pig lymphocyte membranes (Allan et al., 1972) and virus glycoproteins (Hayman et al., 1973).

The interaction of the lectins with the surface structures of cells allows one to separate cell populations. Thus, wheat germ lectin may be used for the fractionation of T lymphocytes into subclasses (Saito et al., 1978) and *Helix pomatia* lectin, for highly purified peripheral T lymphocytes (Hellström et al., 1976). For details, see Chapter 10.

POLYNUCLEIC ACIDS

The biospecific interaction between deoxyribonucleic acids and the complementary ribonucleic acids or their fragments or building blocks is used with great success for affinity chromatographic purification of these substances (Chap. 9). Polyuridylic acid, poly(U), and polyadenylic acid, poly(A) are very suitable group-specific ligands. Supports with these ligands, approximately 100 nucleotides long, which covalently bind to Sepharose 4B by the cyanogen bromide method, are produced by Pharmacia under the trade name Poly(U)-Sepharose 4B and Poly(A)-Sepharose 4B.

Another type of ligands are the derivatives of boric acid, particularly 3-aminobenzene boric acid (Schott et al., 1973). The following reaction underlies this type of ligand. In alkaline medium boric acid and its derivatives form a complex with substances possessing two free cis (coplanar) OH groups, such as the ribose residue in positions 2 and 3, that again dissociates in acidic milieu. Thus it is possible, using boric acid derivatives, to separate mononucleotides, oligonucleotides, and transfer RNA whose ribose residue is unsubstituted in positions 2 and 3 from other accompanying substances.

SPECIAL VARIANTS

In several cases the matrix itself possesses bioaffine properties; that is, structural units or building blocks of the matrix may assume the function of the bioaffine ligand. The best-known examples are the polysaccharide matrices (cross-linked dextran and agarose) for the purification of lectins.

In the 1960s Agrawal and Goldstein (1967) isolated Con A, and Aspberg and coworkers (1968) a lectin from *Vicia crassa* in a one-step procedure by chromatographing the plant extract through a bed of Sephadex. In this procedure lectin is specifically bound to the matrix and may be eluted by a glucose-containing buffer.

Agarose matrix can be used for galactose-specific lectins. To increase the terminal galactose units, Ersson and colleagues (1973) have hydrolyzed the Sepharose matrix without complete degradation of the gel and thus have increased the adsorption capacity of this gel for *Crotalaria juncea* lectin.

Glycoproteins and enzymes may also be purified in this way. Sivakami and Radhakrishnan (1973) reported on purification of rabbit intestinal glucoamylase by specific adsorption on Sephadex G-200 and elution with maltose or starch. The binding of the enzyme appears to be specific to rabbit intestinal glucoamylase.

In a few cases cellulose has been used to purify cellulases (Halliwell and Griffin, 1978; Weber et al., 1980; Nummi et al., 1981). It has been found that well-ordered fibrous cellulose is not accessible to and does not sorb cellulases, but sorption to amorphous regions is rapid.

Other special variants are the application of ligands without a matrix, for example, by immobilization of protein ligands with glutaraldehyde as cross-linking agents (Avrameas and Guilbert, 1971) or by adsorption of ligands to the matrix and then immobilization by cross-linking (Tardy et al., 1978). The last variant may be used on an industrial scale.

REFERENCES

Agrawal, B. B. I., and Goldstein, I. J. (1967). Protein carbohydrate interaction. VI. Isolation of concanavalin A by biospecific adsorption on cross-linked dextran gels. *Biochim. Biophys. Acta* *147*: 262-271.

Allan, D., Auger, J., and Crumpton, M. J. (1972). Glycoprotein receptors for concanavalin A isolated from pig lymphocyte plasma membrane by affinity chromatography in sodium deoxycholate. *Nature New Biol.* *236*: 23-25.

Aspberg, K., and Porath, J. (1970). Group-specific adsorption of glycoproteins. *Acta Chem. Scand.* *24*: 1839-1841.

Aspberg, K., Holmen, H., and Porath, J. (1968). A non-specific phytohemagglutinin found in vicia cracca. *Biochim. Biophys. Acta* *160*: 116-117.

Avrameas, S., and Guilbert, B. (1971). Biologically active water-insoluble protein polymers. Their use for the isolation of specifically interacting proteins. *Biochimie* *53*: 603-614.

Berg, K., Heron, I., and Hamilton, R. (1978). Purification of human interferons by antibody affinity chromatography using highly adsorbed anti-interferon. *Scand. J. Immunol.* *8*: 429-436.

Brodelius, P., and Mosbach, K. (1973). Separation of the isoenzymes of lactate dehydrogenase by affinity chromatography using an immobilized AMP-analogue. *FEBS Lett.* *35*: 223-226.

Brodelius, P., Larsson, P.-O., and Mosbach, K. (1974). The synthesis of three AMP-analogues: N^6-(6-aminohexyl)adenosine-5'-monophosphate, N^6-(6-aminohexyl)adenosine-2',5'-bisphosphate and N^6-(6-aminohexyl)adenosine-3',5'-bisphosphate and their application as general ligands in biospecific affinity chromatography. *Eur. J. Biochem.* *47*: 81-89.

Chua, G. K., and Bushuk, W. (1969). Purification of wheat proteases by affinity chromatography on hemoglobin-Sepharose column. *Biochem. Biophys. Res. Commun.* *37*: 545-550.

Cuatrecasas, P., Wilchek, M., and Anfinsen, C. B. (1968). Selective enzyme purification by affinity chromatography. *Proc. Nat. Acad. Sci. U.S.* *61*: 636-643.

Davey, M. W., Sulkowski, E., and Carter, W. A. (1976). Binding of human fibroblast interferon to concanavalin A-agarose. Involvement of carbohydrate recognition and hydrophobic interaction. *Biochemistry* *15*: 704-713.

Dean, P. D. G., and Lowe, C. R. (1972). Some studies on insolubilized nicotinamide nucleotides. *Biochem. J.* *127*: 11P.

De Maeyer-Guignard, J., Tovey, M. G., Gresser, J., and De Maeyer, E. (1978). Purification of mouse interferon by sequential affinity chromatography on poly(U)- and antibody-agarose columns. *Nature* *271*: 622-625.

Dignam, J. D., and Strobel, H. W. (1977). NADPH-cytochrome P-450 reductase from rat liver: Purification by affinity chromatography and characterization. *Biochemistry* *16*: 1116-1123.

Edy, V. G., Billiau, A., and De Somer, P. (1977). Purification of human fibroblast interferon by zinc chelate affinity chromatography. *J. Biol. Chem. 252*: 5934-5935.

Ersson, B., Aspberg, K., and Porath, J. (1973). The phytohemagglutinin from sunn hemp seeds (crotalaria juncea). Purification by biospecific affinity chromatography. *Biochim. Biophys. Acta 310*: 446-452.

Halliwell, G., and Griffin, M. (1978). Affinity chromatography of the cellulase system of trichoderma koningii. *Biochem. J. 169*: 713-715.

Harvey, M. J., Craven, D. B., Lowe, C. R., and Dean, P. D. G. (1974a). N^6-Immobilized 5'-AMP and NAD^+: Preparations and applications. *Methods Enzymol. 34*: 242-253.

Harvey, M. J., Lowe, C. R., Craven, D. B., and Dean, P. D. G. (1974b). Affinity chromatography on immobilized adenosine-5'-monophosphate. 2. Some parameters relating to the selection and concentration of the immobilized ligand. *Eur. J. Biochem. 41*: 335-340.

Hayman, M. J., Skehel, J. J., and Crumpton, M. J. (1973). Purification of virus glycoproteins by affinity chromatography using lens culinaris phytohaemagglutinin. *FEBS Lett. 29*: 185-188.

Hellström, U., Hammarström, S., Dillner, M.-L., Perlmann, H., and Perlmann, P. (1976). Fractionation of human blood lymphocytes on helix pomatia A hemagglutinin coupled to Sepharose bead. *Scand. J. Immunol. 5 Suppl.* 5: 45-55.

Huang, J. W., Davey, M. W., Hejna, C. J., von Muenchhausen, W., Sulkowski, E., and Carter, W. A. (1974). Selective binding of human interferon to albumin immobilized on agarose. *J. Biol. Chem. 249*: 4665-4667.

Jankowski, W. J., von Muenchhausen, W., Sulkowski, E., and Carter, W. A. (1976). Binding of human interferon to immobilized Cibacron Blue F3G-A. The nature of molecular interaction. *Biochemistry 15*: 5182-5187.

Lamed, R., Levin, Y., and Wilchek, M. (1973). Covalent coupling of nucleotides to agarose for affinity chromatography. *Biochim. Biophys. Acta 304*: 231-235.

Lloyd, K. O. (1970). The preparation of two insoluble forms of the phytohemagglutinin, concanavalin A, and their interaction with polysaccharides and glycoproteins. *Arch. Biochem. Biophys. 137*: 460-468.

Lowe, C. R. (1978). Effect of the nature of the spacer arm in affinity chromatography on immobilized AMP, in *Chromatography of Synthetic and Biological Polymers*, Vol. 2, R. Epton (ed.). Ellis Horwood Ltd., Chichester, pp. 250-257.

Lowe, C. R., and Dean, P. D. G. (1971). Affinity chromatography of enzymes on insolubilized cofactors. *FEBS Lett. 14*: 313-316.

Lowe, C. R., Mosbach, K., and Dean, P. D. G. (1972). Some applications of insolubilized cofactors to the purification of pyridine nucleotide-dependent dehydrogenases. *Biochem. Biophys. Res. Commun. 48*: 1004-1010.

Mizrahi, A., O'Malley, J. A., Carter, W. A., Takatzuki, A., Tamura, G., and Sulkowski, E. (1978). Glycosylation of interferon effects of tunicamycin on human immune interferon. *J. Biol. Chem. 253*: 7612-7615.

Mosbach, K. (1974). AMP and NAD as "general ligands." *Methods Enzymol. 34*: 229-242.

Mosbach, K. (1978a). Immobilized adenine coenzymes in general ligand affinity chromatography and their use as active coenzymes. In *Affinity Chromatography*, O. Hoffmann-Ostenhoff, M. Breitenbach, F. Koller, D. Kraft, and O. Scheiner (eds.). Pergamon Press, Oxford, pp. 55-66.

Mosbach, K. (1978b). Immobilized adenine coenzymes in general ligand affinity chromatography. In *Chromatography of Synthetic and Biological Polymers*, Vol. 2, R. Epton (ed.). Ellis Horwood Ltd., Chichester, pp. 199-230.

Nummi, M., Niku-Paavola, M.-L., Enari, T.-M., and Raunio, V. (1981). Isolation of cellulases by means of biospecific sorption on amorphous cellulose. *Anal. Biochem. 116*: 137-141.

Pommerening, K., Kühn, M., Jung, W., Buttgereit, K., Mohr, P., Štamberg, J., and Beneš, M. (1979). Affinity chromatography of haemoproteins. 1. Synthesis of various imidazole containing matrices and their interaction with haemoglobin. *Int. J. Biol. Macromol. 1*: 79-88.

Saito, M., Toyoshima, S., and Osawa, T. (1978). Isolation of partial characterization of the major sialoglycoprotein of human T-lymphoblastoid cells of a MOLT-4B cell line. *Biochem. J. 175*: 823-831.

Schott, H., Rudloff, E., Schmidt, P., Roychoudhury, R., and Kössel, H. (1973). A dihydroxyboryl-substituted methacrylic polymers for the column chromatographic separation of mononucleotides. *Biochemistry 12*: 932-938.

Secher, D. S., and Burke, D. C. (1980). A monoclonal antibody for large-scale purification of human leukocyte interferon. *Nature 285*: 446-450.

Sivakami, S., and Radhakrishnan, A. N. (1973). Purification of rabbit intestinal glucoamylase by affinity chromatography on Sephadex G-200. *Indian J. Biochem. Biophys. 10*: 283-284.

Tardy, M., Tayot, J.-L., Roumiantzeff, M., and Plan, R. (1978). Immunoaffinity chromatography on derivatives of porous silica beads. Industrial extraction of antitetanus antibodies from placental blood and plasma. In *Chromatography of Synthetic and Biological Polymers*, Vol. 2, R. Epton (ed.). Ellis Horwood Ltd. Chichester, pp. 298-313.

Turková, J., Valentová, O., and Čoupek, J. (1976). Isolation of aminopeptidase from aspergillus flavus. *Biochim. Biophys. Acta* *420*: 309-315.

Weber, M., Foglietti, M. J., and Percheron, F. (1980). Fractionnement d'une preparation cellulasique de trichoderma viride par chromatographie d'affinite sur cellulose reticulee. *J. Chromatogr.* *188*: 377-382.

Yasukochi, Y., and Masters, B. S. S. (1976). Some properties of a detergent-solubilized NADPH-cytochrome c (cytochrome P-450) reductase purified by biospecific affinity chromatography. *J. Biol. Chem.* *251*: 5337-5344.

6
General Consideration of the Adsorption and Elution Step

In the preceding chapters the properties with which a matrix should be endowed to be suited for affinity chromatography, the advantages and disadvantages of the most common types of matrix, the chemical aspects of matrix activation and subsequent ligand fixation, and the requirements that must be met to an effective matrix are discussed. Two further factors, already mentioned earlier in this contribution, have an essential influence on the success of affinity chromatography:

- The strength of interaction between the covalently bound ligand and its partner
- The steric conditions in the microenvironment at the surface of the matrix

In most of the literature these factors are treated empirically, but there are also abundant data on the theoretical basis of affinity chromatography, derived from the equilibrium model, cooperative bonding within the plate theory, or statistical theory. Surveys have been published by Lowe and Dean (1974), Turková (1978), and Scouten (1981). In the following some parameters are discussed in more detail.

INFLUENCE OF COMPLEX STABILITY

The basic reaction in affinity chromatography is the formation of a specific complex SL between the substance to be isolated S and the matrix-bound ligand L according to the equation

$$S + L \rightleftharpoons SL \quad (1)$$

A measure of the strength of the complex is the dissociation constant or equilibrium constant:

$$K = \frac{[S]\ [L]}{[SL]} \quad (2)$$

In a first approximation, several boundary conditions can be derived from Equation (2) (O'Carra et al., 1973; Lowe and Dean, 1974). If S_0 and L_0 are the initial concentrations of the substance or the matrix-bound ligand and $L_0 \gg S_0$, then

$$K = \frac{[S]\ [L]}{[SL]} \quad (3)$$

$$= \frac{[S_0 - SL]\ [L_0 - SL]}{[SL]}$$

$$= \left(\frac{S_0 - L}{SL}\right) L_0$$

The chromatographic distribution coefficient K_d is defined by

$$K_d = \frac{\text{bound substance}}{\text{free substance}} \quad (4)$$

$$= \left(\frac{SL}{S_0 - SL}\right) = \frac{L_0}{K}$$

or in terms of column units:

$$V_e = V_0 + K_d V_0 \quad (5)$$

where V_e and V_0 are the elution volume of the substance and the void volume, respectively. Hence,

$$\frac{V_e}{V_0} = \frac{L_0}{K} + 1 \quad (6)$$

Equations (4) and (6) show that effective affinity chromatography will be achieved only if the condition $L_0 \gg K$ is fulfilled.

The upper limit value of the ligand concentration at the matrix is 10^{-3}-10^{-2} M. Assuming that $V_e/V_0 > 3$ for the separation to be effective, the lower limit for K is 10^{-3}-10^{-4} M. Furthermore, it can be inferred from this consideration that with complexes with $K < 10^{-4}$ M a minimum concentration of the ligand at the matrix is necessary for affinity chromatography.

As shown in Figure 6.1, various elution profiles are possible, theoretically, depending on K (Cuatrecasas, 1972). Figure 6.1a shows the elution profile with an unsubstituted matrix or a completely uneffective ligand. No interaction is detectable. At a dissociation constant $K < 10^{-3}$ M, provided that no other steric conditions are involved, for example, a specific interaction with the affine matrix is demonstrable which is, however, too weak to be sufficient for the separation (Fig. 6.1b, c).

To Figure 6.1d corresponds a K of 10^{-4} to 10^{-3} M. The substance to be purified is not reversibly adsorbed, yet it is so strongly retarded that separation of the substances is possible. If the dissociation constant $K < 10^{-5}$ M, the substance is firmly adsorbed at the matrix and can be eluted only by altering the milieu (Fig. 6.1e). This is the desirable variant in affinity chromatography. The separation is unequivocal and independent of the geometry of the column; the purification may be improved by intensive washing. One should not underestimate the advantage that reversible adsorption can be made from extremely dilute solutions, allowing at the same time concentration by several orders of magnitude. The variant in Figure 6.1d has the advantage that the chromatography requires no alterations in milieu; however, the conditions must be kept in very close limits, and column geometry (e.g., length of the column) plays an important role.

The optimal range for the dissociation constants is 10^{-4} 10^{-8} M in free solution. In a case of very high affinity ($K < 10^{-8}$ M), as in the affinity between a hormone and its receptor, the elution step may be extremely difficult and may, under certain circumstances, proceed only under denaturing conditions.

It should be emphasized that it is not readily possible to infer from the affinity of the partners in solution to their behavior in affinity chromatography. Especially with the fixation of low-molecular-weight ligands, the support matrix may have both an inhibitory and, due to additional interactions, a stimulating effect on complex formation. Thus, the matrix does not play only the passive role of the solid-phase component but may essentially influence the stability of the complex.

Generally, matrix fixation of the one partner entails a decrease of affinity, but the contrary is also observed, as, for instance, with the interaction of Cibacron Blue F3G-A with enzymes (Amicon, 1980). So far our considerations have disregarded the fact that after matrix fixation essential parameters are altered in comparison with a free ligand in solution. The concentration is not uniform in the volume unit, and the microenvironment of a matrix-bound ligand is determined not only by the aqueous milieu but also by the matrix.

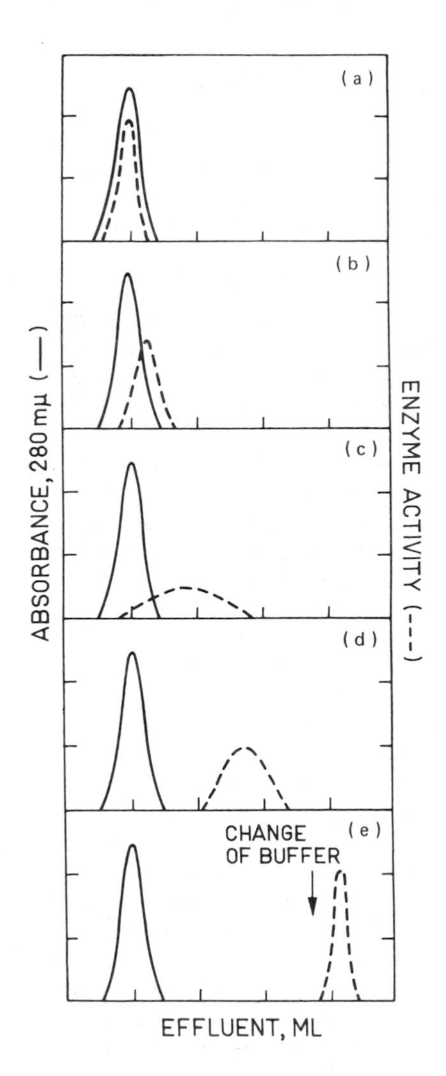

FIGURE 6.1 Theoretical affinity chromatographic patterns obtained by passing a crude protein mixture containing a specific enzyme to be purified (- - - -) on a column containing an adsorbent specific for that enzyme. Successful application is depicted in (e). Various degrees of ineffectiveness are demonstrated in the other column emergence patterns. (From Cuatrecasas, 1972.)

STERIC CONDITIONS

In affinity chromatography the specific complex formation is characterized by the fact that the one partner, owing to its covalent linkage to the matrix, is itself a part of the matrix. Hence there may occur the following steric problems that inhibit or prevent complex formation:

Diffusion limitation
Steric interferences
Blockade of the active site of the ligand through covalent fixation

Matrix fixation not only prevents diffusion of one partner but may also inhibit or prevent interaction of the free partner with the matrix-bound partner because of inappropriate porosity. Therefore, macroporosity is a basic requirement for matrix material to be suitable for affinity chromatography.

Lowe and Dean (1971) have systematically investigated the influence of porosity on the interaction of NAD-Sephadex with serum albumin, lactate dehydrogenase, and malate dehydrogenase. With NAD^+-Sephadex G25 all three proteins appear in the void volume, because the enzymes are unable to diffuse to the matrix-bound ligands. This behavior changes by increasing porosity.

With NAD^+-Sephadex G100, malate dehydrogenase was adsorbed, but the lactate dehydrogenase, with its higher molecular weight, mostly passed through the column. With NAD^+ Sephadex G200 both enzymes were bound to the matrix. This combination of affinity chromatography with gel permeation can be used quite specifically for the separation of two compounds with the same affinity to the bound ligand yet different molecular weights.

Matrix fixation of small ligands can affect the complex formation for steric reasons. This problem becomes especially critical when the active site of a substance is located in a protein pocket. To minimize steric interference and to increase the flexibility and mobility of small ligands, the ligand is bound to the matrix, not directly, but through a spacer arm.

The importance of the spacers has been recognized very early. Cuatrecasas and colleagues (1968) showed in their first studies the high effectivity of immobilized ε-aminohexanoyl-D-tryptophan methyl esters in the purification of α-chymotrypsin compared with the D tryptophan methyl ester bound directly to the matrix. Cuatrecasas (1970) has found at separation of staphylococcal nuclease by an inhibitor [3'-(4-aminophenylphosphoryl)deoxythymidine-5'-phosphate] bound to agarose that the efficiency of this affinity chromatographic method can be considerably enhanced by interposition of chains of different length as it is shown in Figure 6.2.

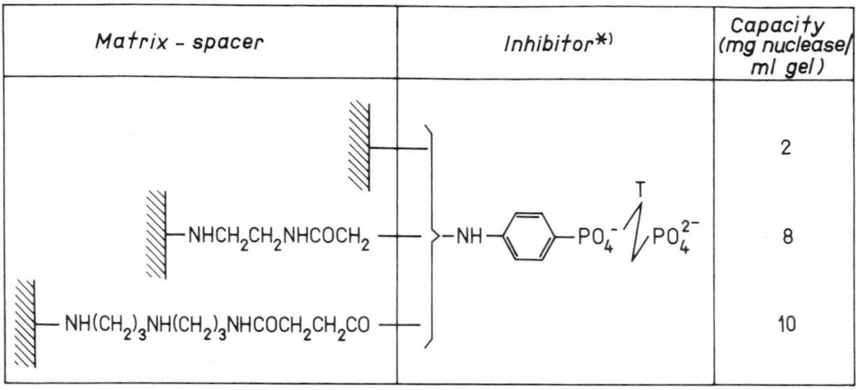

Matrix - spacer	Inhibitor*)	Capacity (mg nuclease/ ml gel)
(matrix) —		2
(matrix) —$NHCH_2CH_2NHCOCH_2$—	—NH—(phenylene)—PO_4^- (T) PO_4^{2-}	8
(matrix) —$NH(CH_2)_3NH(CH_2)_3NHCOCH_2CH_2CO$—		10

*) 3′-(4-aminophenylphosphoryl)-deoxythymidine 5′-phosphate

FIGURE 6.2 Capacity of columns containing various Sepharose 4B adsorbents for staphylococcal nuclease.

In detailed studies the use and nature of the spacers have been reported. Figure 6.3 illustrates schematically the influence of spacer length. If no spacer is interposed or if a spacer is too short, no complex formation can take place; the ligand is ineffective (Fig. 6.3a). With the increase in chain length an interaction occurs that does not, however, correspond to the optimal conditions (Fig. 6.3b). Full interaction is achieved only with sufficient chain length (Fig. 6.3c). This general consideration shows that behavior in affinity chromatography allows conclusions to be drawn on the steric conditions in the active center of a biomacromolecule. If a spacer chain is very long, unspecific effects may predominate in the interaction. These effects are utilized especially in the so-called hydrophobic chromatography (Chap. 15).

Since the structural parameters of the substances to be isolated differ strongly, no generally optimal spacer can be specified with regard to length and chemical nature. In most cases a C_6 chain is sufficient. In a first approximation, four methylene groups give a space variation of $\sim$5 Å, and eight methylene groups, of $\sim$10 Å. A few problems involved are discussed in detail using the concrete example of the interaction between imidazole and methemoglobin (Pommerening et al., 1979). Imidazole and its derivatives form a complex with Fe(III)-protoporphyrin IX (hemin) in free form or as a prosthetic group of methemoglobin. Complex formation is accompanied by characteristic changes of absorption spectra and electron paramagnetic resonance (EPR) (Pommerening et al., 1980).

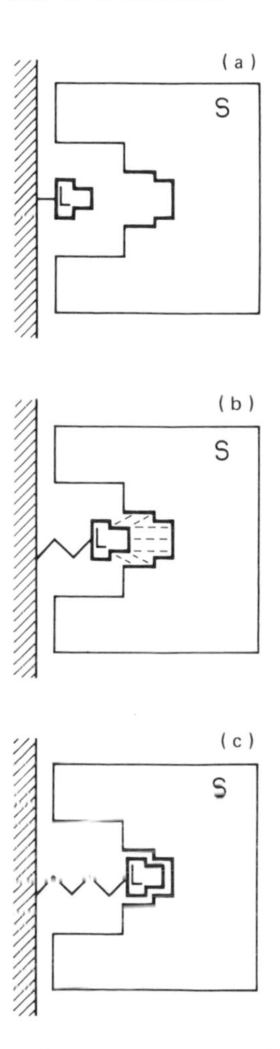

FIGURE 6.3 The principle of function of a spacer arm: (a) nonbinding, (b) low interaction, (c) optimal interaction between ligand (L) and substance (S).

In hemoglobin the heme disk lies in a narrow hydrophobic protein pocket near the surface (Fig. 6.4a). The distance between iron and the protein surface can be assumed to be roughly 5 Å. Of the six coordination positions of iron, four are occupied by the nitrogen of the pyrrole cycles of porphyrin, whereas the fifth interacts with a histidine residue of the protein component. In oxyhemoglobin the sixth position is occupied by oxygen and the iron is present in the reduced

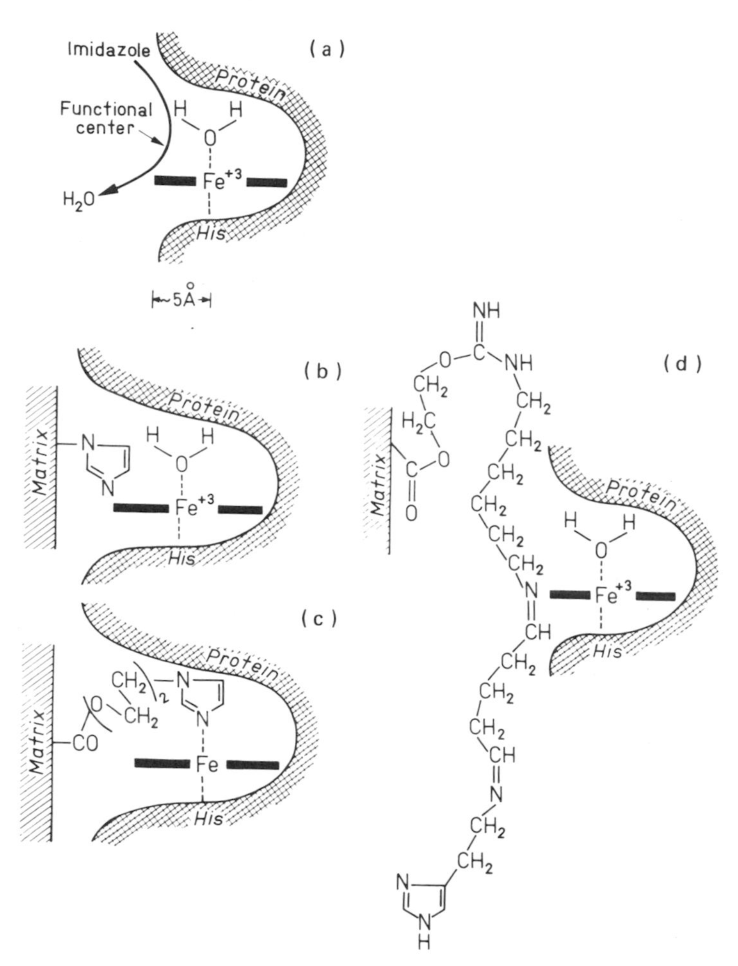

FIGURE 6.4 Interaction of methemoglobin- and imidazole-containing matrices. Dependence on spacer length. (a) functional center of methemoglobin-water complex, (b) without spacer, (c) adapted spacer, (d) long spacer.

Fe^{2+} form. This oxyhemoglobin forms no complex with the imidazole, but the methemoglobin does, where the iron is present in the oxidized form as Fe^{3+} and the sixth position is occupied by water or OH^- in dependence on pH.

Complex formation with matrix-bound imidazole is influenced by the structure of the protein pocket (size, hydrophobicity, distance of iron from the protein surface, and oxidation state of the iron), spacer length and structure, and the interaction between protein and matrix/spacer. It was found that the length and the structure of the spacer, as well as the substituents at the imidazole ring, are of critical importance, and different mechanisms of action must be discussed (Fig. 6.4). With a spacer length of <5 Å, as with poly-N-vinylimidazole or cellulose, whose OH group at the C_6 atom is substituted by imidazole, no interaction occurs for steric reasons, in contrast to free protohemin (Fig. 6.4b). With a spacer length of >5 Å, adsorption takes place by complex formation between imidazole and the prosthetic group independently of the linkage in position 1 or 4(5) of the imidazole ring to the matrix (Fig. 6.4c). The optimal spacer length was found to be n = 1 or 2 (4.8 and 8.7 Å, respectively). The coordinative attachment of the imidazole to the iron of the prosthetic group was evidenced by spectroscopy. The visible spectra and the EPR spectra showed the corresponding maxima and g values of the imidazole complex (Pommerening et al., 1980).

Complex formation is the decisive step but is not solely responsible for the stability of the complex. The complex is so stable that desorption is possible only under denaturing conditions. Hence it is concluded that the methemoglobin-matrix complex was not only stabilized by the imidazole-iron binding but by other interactions, such as the hydrophobic interaction between the two complex partners.

The importance of the structure of the spacer is obvious from the following compilation (Fig. 6.5). Small changes provoke drastic changes in adsorption behavior. Figure 6.5a shows the unmodified side chain of a hydroxyethylmethacrylate support. No binding of hemoglobin is observed. If the imidazole residue is substituted for the OH group (Fig. 6.5b), then a specific binding of methemoglobin occurs, but not of oxyhemoglobin. If an OH group is in the β position from imidazole at the spacer, as occurs with activation by epichlorhydrin (Fig. 6.5c), the adsorption of methemoglobin decreases drastically. If the OH group is replaced by a primary amine, such as ethylamine (Fig. 6.5d), then a hydrophobic interaction occurs irrespective of the oxidation state of the iron.

Large substituents at the imidazole ring disturb the complex formation but not the adsorption of hemoglobin. In this case hydrophobic interactions are responsible. With a long spacer (20 Å) hydrophobic interactions are predominantly responsible for the adsorption (Fig. 6.4d). This could be proved by the following. Imidazole does not play any role in adsorption process. The adsorption is independent of the oxidation state of the iron, and the EPR spectra of adsorbed hemoglobin did not show the corresponding g values of the imidazole complex. This example shows the complexity of the interaction and its dependence on many factors, but it also shows that from the behavior in affinity chromatography conclusions can be drawn on the molecular structure of the binding site.

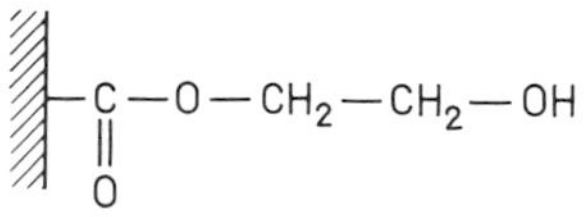

$-C(=O)-O-CH_2-CH_2-OH$ (a)

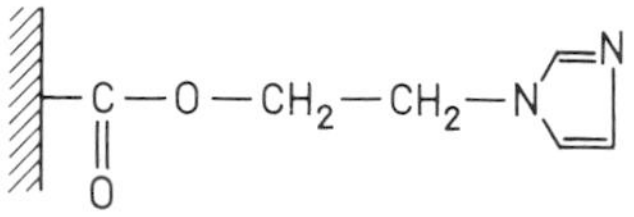

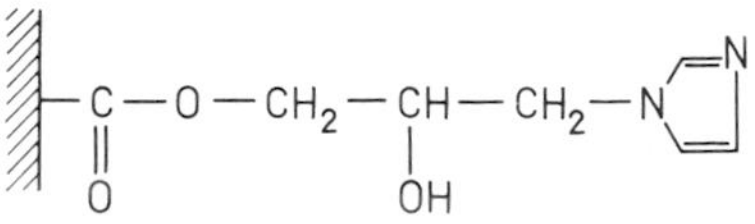

$-C(=O)-O-CH_2-CH_2-N$(imidazole) (b)

$-C(=O)-O-CH_2-CH(OH)-CH_2-N$(imidazole) (c)

$-C(=O)-O-CH_2-CH_2-NH-CH_2-CH_3$ (d)

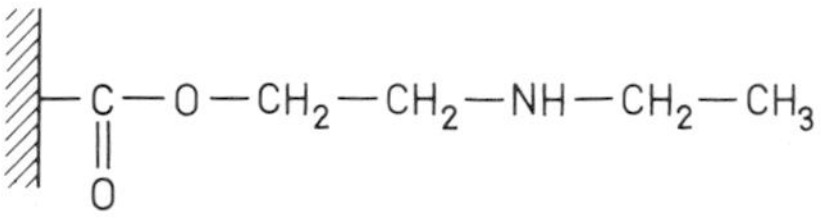

FIGURE 6.5 Side chains of modified hydroxyethylmethacrylate gels: (a) nonmodified (Spheron), (b, c) imidazole containing methacrylate gels, (d) methacrylate gel with hydrophobic side chain.

Covalent fixation of the ligand should occur via functional groups that are not involved in the interaction with the bioaffine partner. Similarly, the matrix fixation must not cause any conformational change or any other intervention into the binding area. The proper choice of the linkage not only affects the effectivity of affinity chromatography, but also permits conclusions on the functional groups involved in the binding. This can be demonstrated by two examples. Various dehydrogenases and kinases show a different behavior toward the two immobilized AMP derivatives (Harvey et al., 1974). In the first matrix AMP is bound to Sepharose via the adenine part in position 6, and in the second matrix AMP is bound by 5'-phosphate (Fig. 6.6). Some enzymes (lactate dehydrogenase, malate dehydrogenase, 3-phosphoglycerate kinase, and pyruvate kinase) are adsorbed to both matrices, whereas other enzymes (hexokinase and creatine kinase) are not adsorbed to either matrix (Table 6.1). Alcohol dehydrogenase and glycerokinase are only bound to N^6-(6-aminohexyl)-AMP-Sepharose and glucose-6-phosphate dehydrogenase, p-glyceraldehyde-3-phosphate dehydrogenase, and myocinase only to P^1-(6-aminohexyl)-P^2-(5'-adenosine)pyrophosphate Sepharose. This shows the difference in the nature of the enzyme nucleotide interactions. In some

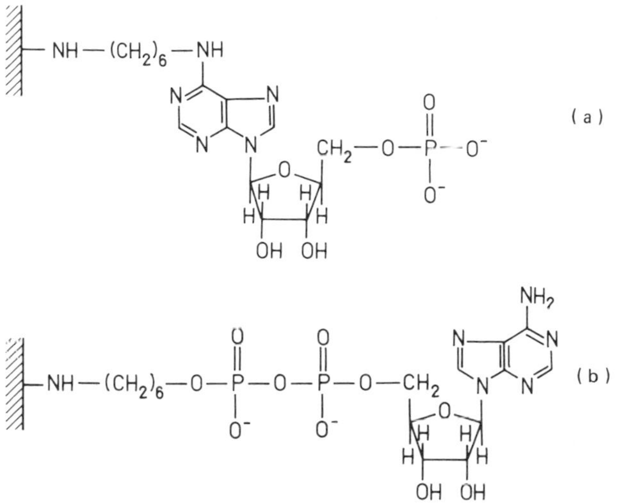

(a)

(b)

FIGURE 6.6 Partial structure of N^6-(6-aminohexyl)-AMP-Sepharose (a) and P^1-(6-aminohexyl)-P^2-(5'-adenosine)pyrophosphate-Sepharose (b).

cases the free 5'-phosphate group or the adenosine part, or both, is essential for the interaction.

Cibacron Blue F3G-A may be coupled to the matrix by different methods (Chap. 12), which give different structures of the dye-matrix. The fixation may take place via the triazine ring or via the primary amino group of the anthraquinone residue. In the case of dextran blue both structures are realized. These three types show different behaviors with regard to the affinity of the proteins. Birkenmeier (1981) found that the specific interaction with serum albumin is lost when matrix fixation of the dye proceeds via the primary amino group. For specific binding of the albumin not only a certain distance of the anthraquinone residue from the matrix but also the free amino group are necessary.

LIGAND CONCENTRATION

The concentration of the immobilized ligand has a distinct influence on the effectivity of a bioaffine support. It is not necessary to prepare a support with a high concentration of the ligand. With increasing ligand density the binding strength and the nonspecific interaction

TABLE 6.1 Difference of the Adsorption of Cofactor-Dependent Enzymes to $\underline{N}^6$-(6-aminohexyl)-5'-AMP-Sepharose (N) and $\underline{P}^1$-(6-aminohexyl)-$\underline{P}^2$-(5'-adenosine)pyrophosphate-Sepharose (P)

Enzyme	Source	Binding (β)[a]	
		N	P
Lactate dehydrogenase	Pig heart, rabbit muscle	>1000[b]	>1000[b]
Malate dehydrogenase	Pig heart	65	490
3-Phosphoglycerate kinase	Yeast	70	260
Pyruvate kinase	Rabbit muscle	100	110
Alcohol dehydrogenase	Yeast	400	0
Glycerokinase	*Candida mycoderma*	122	0
Glucose 6-phosphate dehydrogenase	Yeast	0	170
D-Glyceraldehyde 3-phosphate dehydrogenase	Rabbit muscle	0	>1000[b]
Myokinase	Rabbit muscle	0	380
Hexokinase	Yeast	0	0
Creatine kinase	Rabbit muscle	0	0

[a]Binding (β) refers to a measure of the strength of the enzyme-immobilized nucleotide interaction and is the KCl concentration (mM) at the center of the enzyme peak when the enzyme is eluted with a linear gradient of KCl; 5 U enzyme applied to a column (5 X 50 mm) containing 1 g of the affinity adsorbent; N (1.5 μmol/ml of AMP), P (6.0 μmol/ml of AMP).
[b]Elution was effected by a 200 μl pulse of 5 mM NADH.
Source: Harvey et al. (1974).

increase. At the same time, the effectivity of the matrix decreases owing to steric hindrance. In most cases a ligand concentration up to 20 μmol/ml is sufficient. Only in cases of systems with low affinity ($K > 10^{-4}$ M) should the highest attainable concentration be used. For these systems in particular ligand concentration is a critical parameter, but also in cases with higher affinity the concentration of the ligand at the matrix should be neither too low nor too high. Holroyde and coworkers (1976) found for the affinity chromatography of glucokinase with 2-amino-2-deoxy-D-glucopyranose-N-(6-aminohexyl)-Sepharose, for example, an optimal concentration of the ligand of 3.75 μmol/g. If the concentration is very low (1.2 μmol/g), then the enzyme does not separate from the impurities. At high concentrations (10 μmol/g) the interaction is very high, and drastic conditions are needed to elute the enzyme. It has been found by Turková and colleagues (1982) that the affinity chromatographic efficiency of (ε-aminocaproyl-L-Phe-D-Phe-OCH_3)-Separon, a specific adsorbant for porcine pepsin, depends strongly on the inhibitor concentration. At low concentration (0.85 μmol inhibitor per gram dry gel), pepsin was eluted from the column by increased ionic strength of the elution buffer in a single sharp peak, whereas at high inhibitor concentration (155 μmol/g dry carrier) several peaks of pepsin were observed in the elution diagram exhibiting the same proteolytic activity (Fig. 6.7). These results are attributed to additional unspecific multipoint ligand-enzyme interactions (e.g., electrostatic or hydrophobic interactions) with increasing ligand density on the matrix surface. In conformity with this it was found that the inhibitor concentration determined from the working capacity was much lower than that of the inhibitor concentration determined from the amino acid analysis of hydrolysates of the sorbent. As a result it follows that, for biospecific affinity chromatography of porcine pepsin, a sorbent material with the lowest concentration of bound inhibitor would be most efficient. A very high ligand concentration can lead to complications with regard to the specificity of the adsorption and the native elution. If in this case the ligand is charged or possesses hydrophobic residues, these unspecific interactions may override the specific interactions. This is distinctly pronounced with the dye-ligands. Birkenmeier (1981) found that at pH 7 Cibacron Blue F3G-A Sephadex G100 with a ligand concentration of 4 μg/mg matrix adsorbs specifically albumin from human serum. With increasing ligand concentration the adsorption of the other proteins increases, and only a few serum proteins are not bound to a gel with a ligand concentration of 214 μg/g matrix.

The elution of adsorbed macromolecules from highly substituted gel or from gel with a ligand of very high affinity is often possible only under denaturing conditions. In this case the decrease of ligand concentration is the best way for a native elution. This can be achieved very simply by mixing the bioaffine matrix with a corresponding amount of unsubstituted matrix material.

The affinity of the partners and the ligand concentration often have an influence on the column geometry, too. If both are very high, then the column length is not important. Adsorption takes place in the head of the column, and dilute solution can thus be concentrated simultaneously. For systems with low binding capacity the column length is of decisive importance.

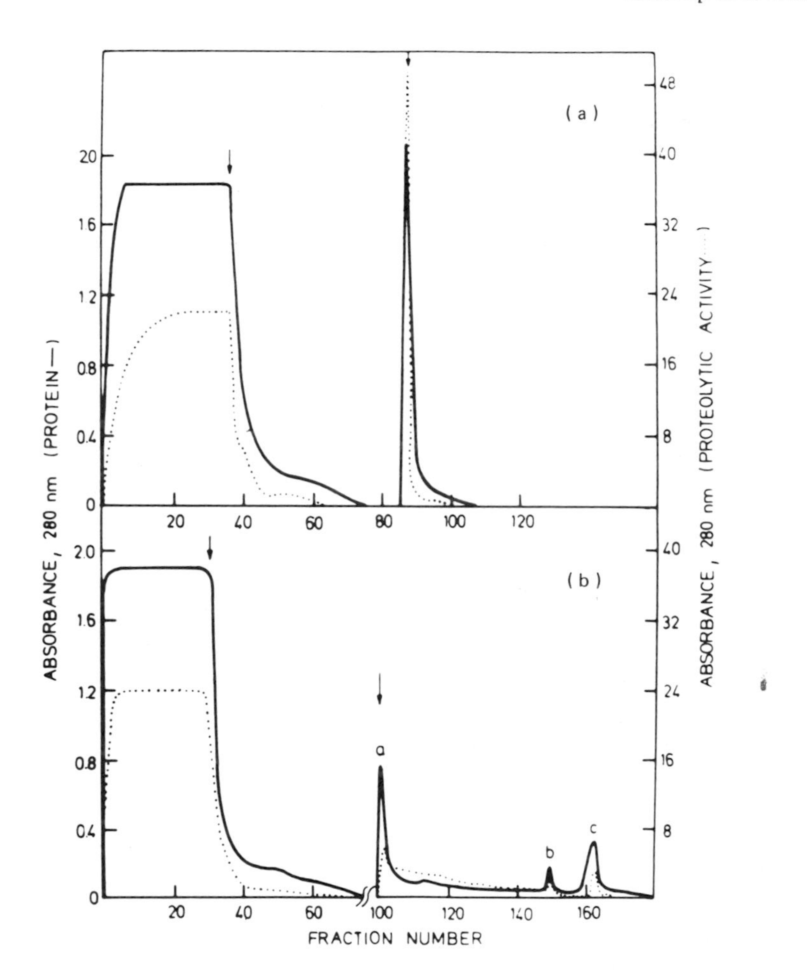

FIGURE 6.7 Affinity chromatography of porcine pepsin on ε-aminocaproyl-L-Phe-D-Phe-OCH_3-Separon columns with (a) low and (b) high concentrations of immobilized inhibitor. The solution of crude porcine pepsin (1 g per 200 ml of 0.1 M acetate buffer, pH 4.5) was applied continuously on the affinity columns (5 ml) equilibrated with 0.1 M sodium acetate (pH 4.5). At the position marked by the first arrow, equilibrated buffer was applied to the columns to remove unbound pepsin and nonspecifically adsorbed proteins. At the second arrow, 0.1 M sodium acetate containing 1 M sodium chloride (pH 4.5) was applied. Fractions (5 ml) were taken at 4 min intervals. The inhibitor concentration of affinity sorbents were (a) 0.85 and (b) 155 μmol/g of dry support. Solid line, protein; broken line, proteolytic activity; a, b, and c, fractions of pepsin of the same specific proteolytic activity. (From Turková et al., 1982.)

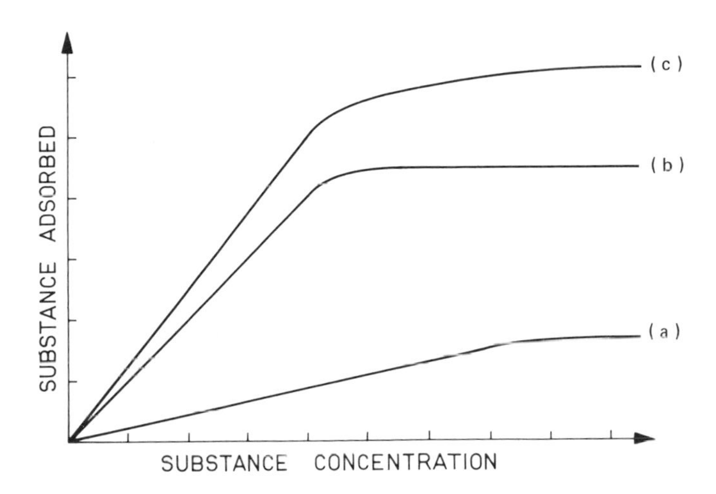

FIGURE 6.8 Adsorption isotherms for affinity chromatography: (a) nonspecific adsorption, (b) ideal biospecific adsorption, (c) experimentally found adsorption.

NONSPECIFIC INTERACTION

The effectivity and selectivity of an affinity chromatography are decreased mainly by nonspecific sorption; this can be demonstrated by the adsorption isotherm (Fig. 6.8). The isotherm found experimentally consists of the specific or the ideal and the nonspecific isotherm. For complex formation of an enzyme, for example, the following equation holds:

$$E + L \rightleftharpoons EL \qquad K = \frac{[E]\ [L]}{[EL]} \qquad \Delta G = -RT \ln K \tag{7}$$

where E is the enzyme, L the matrix-bound ligand, K the equilibrium constant, and ΔG the adsorption energy, assumed to be constant for all adsorbed molecules. The ΔG also consists of the sum of the specific (ΔG_{spec}) and the nonspecific ($\Delta G_{nonspec}$) adsorption energy. Assuming for K a value of 10^{-5} mol/liter, one obtains for ΔG_{spec} a value of about 7 kcal/mol (Turková, 1978). Therefore, $\Delta G_{nonspec}$ should be very low. Nonspecific adsorption depends on many factors and rests largely on hydrophobic and/or ionic interactions; the reasons may be looked for in the matrix itself and in the type of matrix binding, as well as in the nature of the bound ligand. Normally, the hydrophilic matrix materials, for example polysaccharides, have a negligibly low nonspecific interaction.

Nonspecific interaction may distinctly increase after modification. The cross-linking reaction alone increases the nonspecific interaction. The effect becomes very pronounced by activation and immobilization of a ligand. Thus, Jost and coworkers (1974) found that matrix fixation of alkylamines and arylamines by means of the cyanogen bromide method leads to incorporation of ionic groups with an apparent pK value of about 10 for the basic amidine nitrogen.

The increase in nonspecific interaction by incorporation of spacers has been discussed earlier (Chap. 4). The superposition (overlapping) of specific and nonspecific interactions is demonstrated in adsorption experiments using metHb (methemoglobin) and HSA (human serum albumin) in imidazole-containing matrices (Pommerening et al., 1979). Polymethacrylate supports with different imidazole contents were produced via the activated sulfoester. In only one part of the supports were the excessive apolaric sulfoester groups subsequently saponified (Fig. 6.9a). In this case the specific adsorption of methemoglobin is proportional to the imidazole content; the nonspecific adsorption of HSA is low and independent of the imidazole content. The nonsaponified sulfoester groups strongly affect the binding of the two proteins (Fig. 6.9b). Both proteins are adsorbed even at low concentrations of imidazole (high concentration of sulfoester). With increasing imidazole density (decrease of sulfoester concentration) the specific interaction with methemoglobin increases, whereas the nonspecific adsorption of human serum albumin decreases.

PHYSICOCHEMICAL ASPECTS OF AFFINITY CHROMATOGRAPHY

Bernd Ebert and Thomas Hanke, *Academy of Sciences of GDR*

Considerable progress has been made in the application of affinity chromatographic procedures. However, knowledge about the nature and the complex mechanism of the binding process is not on a corresponding level. The reason for this lack of progress was explained recently by Lyklema (1982), who reviewed the forces responsible for the binding between biopolymers and ligands. He stated the reason is primarily that this binding is, as a rule, the compound result of a number of simultaneously occurring molecular processes. This is undoubtedly true, but further problems must be considered. First, for experimental studies of this type of system one must modify and develop new methods, allowing studies on mostly nontransparent and heterogeneous systems. Second, it is difficult, if not impossible, to extend results obtained for a particular case to another one. Third, the occurrence of different molecular processes does not mean that they run down simultaneously, and the sequence and the velocity of these processes will influence the pathway and the effectiveness of the binding process. Nevertheless, there are some examples of special

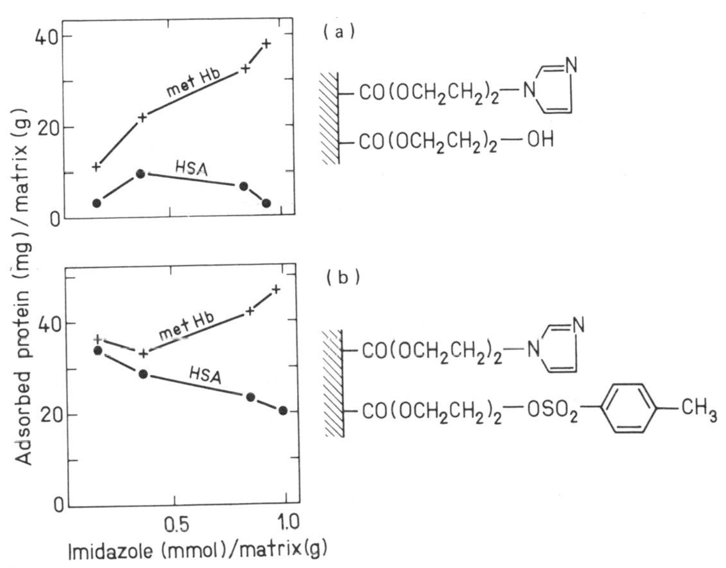

FIGURE 6.9 Adsorption of human methemoglobin (metHb) and human serum albumin (HSA) to polymethacrylate matrices in dependence on imidazole and sulfoester ligands. Matrix (50 mg), 1 ml of 0.8% protein solution, 0.1 M borate buffer, pH 9, containing 0.5 M NaCl, 2 hr incubation, room temperature. Matrix a prepared by saponification of matrix b.

ligands and carriers developed on the basis of general experience. From the third statement it follows that the study of elementary steps, the interaction with substrates and cofactors of the overall reaction in enzyme catalysis, is of general interest with respect to affinity chromatography.

The existence of an enzyme-substrate complex was first demonstrated by Chance (1943) using a stopped-flow apparatus. The conformation of proteins or other biopolymers is a result of a multitude of weak, noncovalent interactions among different parts of the polymer chain and can be influenced by the surrounding medium (pH and ionic strength) or ligands.

The aim of this section is to demonstrate briefly the binding forces on a molecular level and to make an attempt to use spectroscopic methods, mainly electron paramagnetic resonance, to characterize complex systems formed in the adsorption step.

A common way to describe the binding of a given molecule to another is the use of the equilibrium constant K. This term is often applied by chemists and biochemists for the characterization of chemical equilibria. The more general description of the binding process is done in terms of standard Gibbs energies ΔG^o, that is, the isothermal reversible work from the free to the bound state, and vice versa. Specific binding presupposes not only a statistical collision between the molecules and the binding matrix but demands the right orientation. The reorientation (rotation) is temperature dependent and favors the description of the binding in thermodynamic quantities such as ΔG^o.

The Gibbs energy is related to K if the process is reversible; also, ΔG^o contains contributions due to conformational changes (e.g., temporary changes). The reversibility of binding is the basis for affinity chromatography because the bound molecules must be liberated to their original state. To release the bound molecules from the matrix, changes in ionic strength, pH, or solvent are necessary.

Noncovalent binding is mainly based on coulombic forces, hydrogen bonds, van der Waals interactions, and hydrophobic interactions. General procedures for making coulombic as well as van der Waals interactions attractive or repulsive are discussed by Van Oss and colleagues (1980, 1982). These authors distinguish between affinity chromatography and hydrophobic chromatography. This distinction might be plausible, but in most practical cases all possible forces are included. One problem in this consideration is that hydrophobic forces are not well understood and are sometimes described as London-van der Waals forces.

In reality, varying proportions of forces, such as of coulombic or a van der Waals nature, are involved in the interaction between biopolymers and ligands (Van Oss et al., 1982). The coulombic force $f_c(r)$ between two charged molecules (q_1 and q_2, sign included) can be vectorially written as

$$\vec{f}_c(r) = \frac{q_1 q_2 \vec{r}}{4\pi\, \varepsilon\, \varepsilon_0\, r^3} \tag{8}$$

where ε is the dielectric permittivity of the system, $\varepsilon_0 = 8.85 \times 10^{-12}$ $C^2N^{-1}m^{-2}$ and r the distance between two isolated molecules. Subsequently the Coulomb energy U(r) can be written as

$$U(r) = \frac{q_1 q_2}{4\pi \varepsilon \varepsilon_0 r} \tag{9}$$

Another expression for the force f(r) is given by the Gibbs energy G(r) (Lyklema, 1982).

$$f(r) = -\frac{dG(r)}{dr} = -\frac{dH(r)}{dr} + T\,\frac{dS(r)}{dr} \tag{10}$$

The maximum of the function G(r) at given r corresponds to the equilibrium state. In this case it is f(r) = 0 and the situation can be described with equilibrium constants. The determination of equilibrium and rate constants with the help of affinity chromatography has been discussed by Hethcote and DeLisi (1982), for example. The relationship between standard Gibbs energy ΔG^o and the equilibrium constant K is the following:

$$\frac{\Delta G^o}{RT} = pK = -\ln K \tag{11}$$

where the concentrations of the components should be expressed in mole fractions.

A general analysis of electrolyte and solvent effects on the association and interaction of proteins and nucleic acids was given by Record and coworkers (1978). Analyzing the formation of noncovalent macromolecular complexes, the obtained results corresponding to the preceding discussion.

Noncovalent Interactions

In this section different interactions and the approximate stabilization energy are considered. An estimation of stabilization energies has been given by Frieden (1975). The heat of formation of a single covalent bond is about 400 kJ/mol, that is, one order of magnitude higher than the values for noncovalent bonds. The different hydrogen bonds between peptides, between neutral groups, and between neutral and charged groups require 5-40 kJ/mol.

The energy for hydrophobic interactions between methyl groups is 1 kJ/mol and for the stacking of aromatic rings about 5 kJ/mol. The stabilization energy of the interaction of coulombic type (<40 kJ/mol) between different charged groups is attractive and for similarly charged groups repulsive.

The effect of noncovalent interactions arises from their multiplicity and, in some cases, additivity. A large number of weak bonds provides the flexibility and realizes the specificity of biological interaction, as discussed by Frieden (1975). In the review of Lyklema (1982), six types of attractive forces are summarized: coulombic, hydrogen bridges, hydrophobic interaction, and van der Waals forces, including London-van der Waals, Keesom-van der Waals, and Debye-van der Waals forces. One must consider that these forces can be attractive as well as repulsive, excluding the hydrogen bridges and hydrophobic interactions. Additionally, the attractive forces are balanced by the Born repulsion due to the overlap of the outer electron shells. Special attention is paid to long-range forces because they play an important role in adhesion, protein adsorption, and affinity chromatography.

Coulombic interaction depends on the charges of the molecules but not on their specific shape. Coulombic energy depends on r^{-1}; this means that the corresponding force is of long-range type. This force is classified as a mechanical force because it does not depend on temperature.

Recently, Israelachvili and Pashley (1982) stated that hydrophobic long-range interactions are possible and decay exponentially with distance. These authors measured the total force as a function of the distance between two hydrophobic surfaces immersed in aqueous electrolyte solutions. The hydrophobic bonding occurs only in polar media and is not possible between molecules in the gas phase (Franks, 1975). Hydrophobic interactions are commonly described for systems with water as solvent. Water forms, for example, a number of additional hydrogen bonds around hydrocarbon chains dissolved in it. The increasing number of hydrogen bonds leads to a lower mobility and consequently to a lower entropy. The dissolution of hydrocarbons is therefore entropically unfavorable. The association of apolar molecules or domains with each other lowers the entropy because water is liberated from a weakly immobilized state.

Most discussed among noncovalent interactions between biopolymers and ligands and between biopolymers themselves are hydrogen bonds. These bonds are brought about by the interaction of a hydrogen atom between two electronegative atoms, especially nitrogen, oxygen, or sulfur. One group is the proton donor and the other one the acceptor, which is a lone pair or the π-electron orbital of an unsaturated bond. New limiting values of the hydrogen bond geometry were deduced by Höhne and Kretschmer (1982) from crystal structures of biologically active compounds. Geometric parameters of hydrogen bonds in crystal structures were experimentally determined with high accuracy, and the π-electron density of the carbonyl groups as hydrogen bond acceptor in C=0 . . . H-N bonds was found to have an essential influence. A linear arrangement of the atoms in the hydrogen bond mentioned before was not observed in the more than 100 structures considered. This bond holds for $85° < C=0 \ldots H\text{-}N < 165°$, and the distance where a hydrogen bond is reasonable lies between 1.4 and 2.6 Å. These data are qualitatively coincident with data obtained by Jeffrey (1982), who used mainly neutron diffraction measurements. Further consequences of these results are discussed by Höhne and Kretschmer (1984).

These results explain the variability of structures fixed by hydrogen bonds. The importance of hydrogen bonds is supported by the fact that the strongest van der Waals force is in the range of the weakest hydrogen bond. The hydrogen bonds are of short-range type and remain effective up to a distance of about 2.6 Å for the O . . . H bond.

The van der Waals forces can be divided into three different groups: London-van der Waals, Keesom-van der Waals, and Debye-van der Waals.

van der Waals forces are attractive forces between uncharged molecules, and their energy decreases with r^{-6}; that is, they are of short range. This general consideration concerning the short-range action is valid only for two small molecules. The London-van der Waals forces are additive, and the situation changes if the attraction between a biomacromolecule and a ligand is considered. The ligand gives no significant contribution to the long-range action because the number of atoms is small and the influence decays with r^{-6}. The macromolecule, however, has a semi-infinite number of atoms and the energy decays with the third power. This lower decrease is the reason London-van der Waals forces acquire a long range. This type of force describing the interaction of two uncharged molecules behaves like fast oscillating dipoles.

The Keesom-van der Waals formalism has been developed for the case in which two ideal dipoles are interacting. The interaction of dipoles depends on their orientation. Therefore, the thermal motion counteracts with kT the mutual orientation of the two dipoles. The dipoles are usually not fixed so that they can change from an attractive to a repulsive orientation. The statistical averaging of this process leads to an overall attraction because attractive positions are energetically more favorable than repulsive.

Debye-van der Waals forces describe the third possible situation, the interaction between a permanent dipole and an apolar molecule. The basis of the attraction in this system is the polarizability of an apolar molecule, if it is brought into an electrical field. For affinity chromatography it is important to note that Debye forces increase strongly with increasing volume of the apolar molecule. This means that biopolymers adsorbed on polar surfaces show the tendency to turn the most bulky groups toward the surface.

Spectroscopic Investigation

From the rough description of binding forces it becomes clear that different parameters are needed and, therefore, different methods must be used to understand the basic principles of affinity chromatography (including, for example, hydrophobic chromatography) and to optimize the working conditions. Methods that could describe physical properties of the system in this case are, nuclear magnetic resonance, fluorescence depolarization, EPR, and others (Ebert et al., 1982). The advantage of these methods, over radioactive labeling, for example, is that the added or intrinsic probes used show different spectra in the attached and free state, so that with the help of computer techniques equilibrium constants can be determined. The disadvantage consists in the possible disturbing effect of the added probes.

The binding of biopolymers (enzymes) or ligands to polymeric matrices, however, can alter their physical and chemical properties.

One can expect that the immobilization of or the contact with apolaric matrices induces reversible or irreversible conformational changes that influence functional properties. For the optimal utilization of immobilized systems it is of interest to know in what manner the immobilization influences mobility and conformation.

The use of the EPR spin labeling and probe technique is now well established, and it is a method of choice to study heterogeneous nontransparent systems. Some examples for the application of EPR to study different systems were reported by Berliner (1979). From EPR spectra of nitroxide radicals, a rotational correlation time τ can be calculated. From Debye's formula

$$\tau = \frac{V\eta}{kT} \tag{12}$$

where V is the effective volume of rotation, η viscosity, k the Boltzmann constant, and T the absolute temperature, it follows that the rotational correlation time is proportional to a system-dependent viscosity.

In control experiments with capillary viscosimeters and in EPR model studies performed by Ebert and colleagues (1981) and Kuznetsov and Ebert (1974) it has been shown that Debye's formula fits the experimental data if the EPR probe molecules used to determine η values via τ values are one order of magnitude larger in size than the solvent molecules. In this case the macroviscosity can be determined. However, if the probes are comparable in size or even smaller than the solvent molecules, then the formula is no longer satisfied and a new term, microviscosity, is determined.

This does mean in practice that one must add an additional proportionality constant that is system dependent. This latter situation is met in affinity chromatography where small solvent molecules and biopolymers are moving in contact with macromolecular carriers.

The microviscosity in a biopolymer solution reflects the average mobility of the solvent molecules between matrix segments. The dynamics of nitroxide radicals has been studied by Martini and coworkers (1983), for example, who investigated a nitroxide radical adsorbed on porous supports. Different probes and different biopolymer systems lead to different τ values even in pure solvent. Consequently, microviscosity must be determined as a relative quantity comparing the probe in pure solvent and in the matrix-biopolymer systems.

The situation is different if a spin label is covalently attached at a specific site of the protein or on the polymer matrix. In this case the label can reflect changes in the local mobility and polarity, as was shown by Lassmann and colleagues (1973). With the help of this technique the local mobility and the polarity of a Sephadex G200 matrix was analyzed in the water-insoluble gel state as well as in the partly degraded water-soluble state (Pommerening et al., unpublished). For this purpose, the dialdehyde groups of the Sephadex were labeled with 2,2,5,5-tetramethyl-3-aminopyrrolidine-1-oxyl. The rotational correlation time calculated from the obtained spectra was in both cases for swollen dialdehyde Sephadex-gel and after alkaline degradation of the gel matrix eight times lower than calculated for the free label under the same condition. This means that the mobility of the Sephadex chains is in a certain range of molecular weights independent of the overall chain length and the corresponding conformational state. The environment of the NO group of the nitroxide label is in both states strongly polar. Matrix-bound solubilized human hemoglobin was characterized by Pommerening and colleagues (1973). Subsequently, hemoglobin bound covalently to Sephadex was examined by Lassmann and coworkers (1974). Hemoglobin was labeled at the β-93-SH position, which is located near the contact area between $\alpha_1\beta_2$ subunits and therefore reflects the changes of the quarternary structure. The authors were able to show that the structure remains unchanged and that the rotational mobility of hemoglobin decreases 3.5 times after matrix binding.

The importance of continuous flow and stopped-flow methods has already been mentioned. For the study of reactions between biopolymers and ligands these methods were developed mainly in connection with such optical methods as optical rotation, absorbance, and circular dichroism and have been adapted to magnetic resonance methods, like EPR, in a few cases only (Kertesz and Wolf, 1973; Klimes et al., 1980; Ohno, 1982). These techniques allow the determination of rate constants after rapid mixing up to milliseconds.

Nuclear magnetic resonance zeugmatography (tomography) has become a known method for tomographic investigation in medicine. A corresponding EPR method with a stationary field gradient was described by Karthe and Wehrsdorfer (1979).

This technique was improved by Herrling and colleagues (1982) using a modulated field gradient. This technique allows the investigation of the spatial distribution of paramagnetic centers in a sample with a volume of about 1 cm^3. The method can be applied to studies of ligand interactions in heterogeneous nontransparent systems.

Various questions can be investigated in future, for example, the distribution of naturally occurring paramagnetic centers in tissue slices or carrier systems and the distribution of spin probes and their diffusion profiles in dependence on time. This method could be especially useful to study high polymers and catalysts, paramagnetic molecules adsorbed on interfaces, and substrate radicals formed in the system under investigation (Mohr et al., 1983). With the continuous development of instrumentation one can expect a further increase of the application of spectroscopic methods to affinity chromatography.

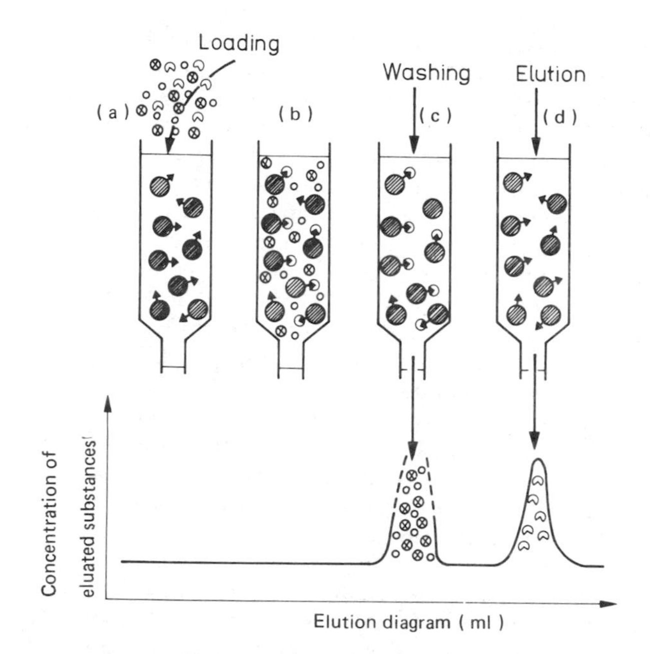

FIGURE 6.10 Column procedure of an affinity chromatography. (a) The column is filled with the matrix-bound ligand ( ) and loaded with the sample containing the bioaffine component and other substances ( ). (b) The bioaffine component contacts the ligand immobilized to the matrix particles ( ). (c) The nonbound substances are removed by washing, and in the elution diagram a more or less sharp peak appears. (d) The bioaffine component is eluted by change of pH, ionic strength or addition of detergents or protein unfolding agents.

PRACTICAL ASPECTS OF AFFINITY CHROMATOGRAPHY

Biospecific complex formation involves various kinds of binding: ionic interaction, hydrogen bonds, hydrophobic interactions, van der Waals-London forces, and others. This is the reason for the frequently drastic dependence of the affinity of the two partners on milieu. This is also reflected by the fact that biologically active substances show a milieu-dependent optimum of action. Well known is the pH function of enzymatic activity. This dependence of complex formation on external conditions is utilized in both the adsorption and the elution step. In Figures 6.10 and 6.11 a general scheme of affinity chromatography in preparative scale is shown.

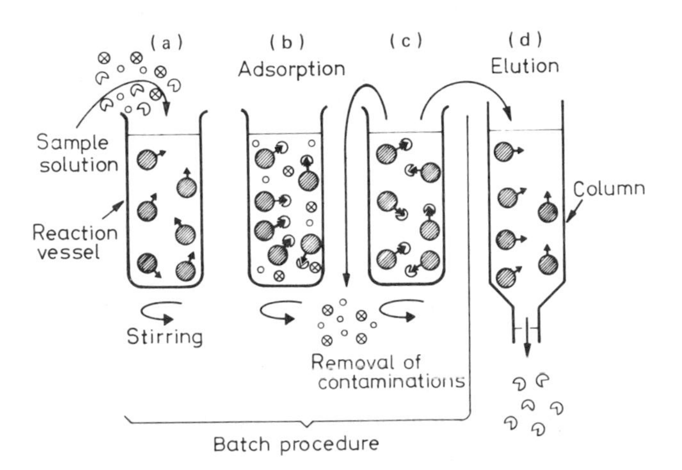

FIGURE 6.11 Batch procedure of an affinity chromatography. (a) The sample solution is given in a reactor vessel. (b) With stirring, the bioaffine component contacts the immobilized ligand. (c) The contaminations are removed by washing, decantation, centrifugation, or filtration. (d) The matrix is given in a filtering equipment or column. The elution takes place as described for the column procedure (Fig. 6.10).

Conditions of the Adsorption Step

Adsorption should take place under conditions ensuring maximal complex formation. Although in a first approximation the optimal adsorption conditions are compatible with physiological conditions, distinct deviations may occur because the partner is bound to a matrix. There are various reasons for this, which have been discussed in the preceding chapters or will be outlined in the following chapters using concrete examples. Therefore it is necessary to determine by preliminary experiments, above all, the pH optimum or the optimal ionic strength for the adsorption.

Because of the temperature sensitivity of many biomacromolecules it is not only desirable but in many cases an urgent necessity to perform affinity chromatography at 2-4°C, as are the other steps in the preparational procedure. The temperature, however, has a clear influence on complex stability. The adsorption is exothermic. The more exothermic the interaction, the more temperature sensitive is the adsorption. Generally, under chromatographic conditions a decrease of temperature will shift the equilibrium to lower concentrations in the mobile phase; that is, low temperatures increase the adsorption. But opposite examples are also known, above all if hydrophobic interactions are

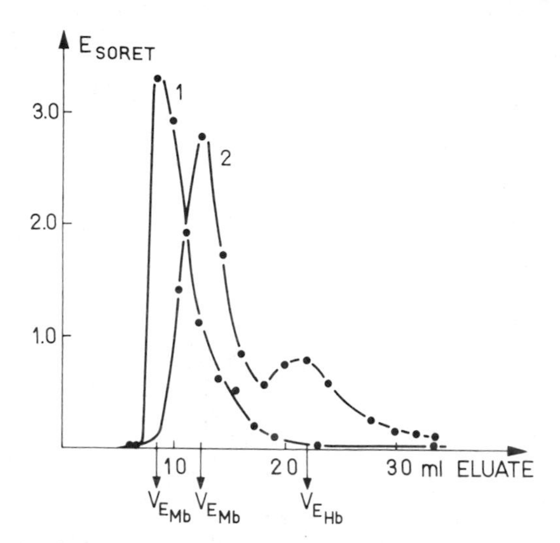

FIGURE 6.12 Separation of a metmylglobin-methemoglobin (bovine) mixture with a Spheron matrix and the following side chains

$$(-\underset{\substack{\| \\ O}}{C}-O-CH_2-CH_2-O-\underset{\substack{\| \\ NH}}{C}-NH-(CH_2)_6-N{=}CH-(CH_2)_3-CH{=}N-CH_2-CH_2-\text{(imidazolyl, N / N–H)}):$$

(1) elution diagram at room temperature, (2) elution diagram at 4°C, 0.1 M borate buffer, pH 8.

prevalent. Figure 6.12 gives an example (Pommerening, unpublished). Earlier in this chapter it was described that the interaction between methemoglobin and imidazole containing matrixes with a very long side chain occurs exclusively via hydrophobic interactions. If a myoglobin-hemoglobin (bovine) mixture is applied at pH8 on a column with a Spheron matrix and this side chain, one curve 1 obtains at room temperature. The methemoglobin remains adsorbed at the matrix quantitatively, and myoglobin elutes with a sharp peak. However, at 4°C (curve 2), the interaction of methemoglobin with the matrix is so weak that it is being eluted, but with a distinct delay as compared with myoglobin. Since $V_{Mb} < V_{Hb}$, a molecular sieving effect is out of the question for the separation.

In the adsorption step the ionic strength should be as low as possible. To suppress nonspecific adsorptions it may be advantageous to work at high ionic strength, such as in dye-ligand chromatography in the presence of 0.5 M NaCl (Chap. 12). The importance of ionic strength in hydrophobic interaction chromatography is discussed in Chapter 15 in more detail.

The adjustment of the equilibrium between the solid and the mobile phase is a time-dependent process. Distinct differences are obvious here between the column and the batch process. Generally, the equilibrium is achieved more quickly in the column procedure. Nevertheless, the lowest acceptable flow rate should be used, if possible not higher than 10 ml cm^{-2} hr^{-1}.

If the volume of sample is smaller than the column volume, after application of the sample one can stop chromatography to facilitate equilibrium adjustment and run the washing and elution processes a few hours later. Nonbound quantities of the substance are frequently due to inappropriate handling. The rapid flow rate is not always to blame for this; the reason may be also in a too-high initial concentration. The process is known as secondary exclusion. If a concentrated solution of biomacromolecules is applied to a porous matrix, then some biomacromolecules cannot diffuse into the pores because these are already occupied by other molecules of the same kind. This leads to partial exclusion from the pores. The amount of the adsorbent is determined by its binding capacity and the amount of the substance to be purified. The latter should be dissolved in the starting buffer, with the volume of high-affinity systems of secondary importance. After adsorption the column should be washed with the starting buffer (10 column volume) followed by the elution step. Only when using weakly bound systems should the substance to be purified be applied in a small volume (5% of the column volume) for the separation to be quantitative.

The batch procedure has an advantage for large-scale isolation with ligands possessing high affinity. The adsorption process can be advantageously performed by the batch procedure, but the elution is best done by the column procedure.

Conditions of the Elution Step

Specific adsorption is only one step for successful affinity chromatography. Equally important is the subsequent native elution. There are many examples in the literature showing that an affinity chromatographic system could not be used because the elution step was impossible to realize. For instance, matrix-bound hemoglobin can bind haptoglobin generally, but it was not possible in all cases to elute the haptoglobin (Pommerening et al., 1978).

In contrast to the adsorption step, in the elution step it is attempted to weaken the interactions of both complex partners by milieu variation to achieve desorption of the column. For the elution various methods have been developed that are either nonspecific or specific in nature (Table 6.2). They may be used partly in combination with the various elution variants (stepwise, gradient, or pulse elution). In many cases a change of pH leads to desorption, especially if complex binding proceeds predominantly through ionic interactions. The pH differences, of course, are limited by the chemical stability of the substance, the matrix, and the covalent linkage between matrix and ligand.

TABLE 6.2 Summary of Elution Methods

Nonspecific methods	Change of pH, ionic strength, temperature, polarity, structure (deforming eluents: urea, guanidine, detergents, chaotropic ions)
Specific method	Affinity elution (with substrates, cofactors, inhibitors, and other specific soluble substances)
Special methods	Electrophoresis, cleavage of matrix-ligand bond, buffer effects

Nevertheless, it can be rather drastic; in the case of proteases and inhibitors (Turková et al., 1973) adsorption proceeds at pH 8.1 and elution at pH 3.1, or with immunoglobulin complexes adsorption occurs at pH 7 and elution between pH 2.4 and 2.8. The increase of ionic strength (gradient or stepwise) is one of the most gentle methods of unspecific elution. Sodium or potassium chloride is mostly used at concentrations between 0.1 and 2 M. In a case of high affinity, chaotropic salts (SCN^- or CCl_3COO^-) can be used, particularly in immunoaffinity chromatography (Chap. 8), or for plasma protein isolation (Chap. 12). In some cases decrease of polarity, by addition of ethanol or dioxan or as with human lymphoid interferon, up to 50% ethylene glycol (Erickson and Paucker, 1979) is used for desorption. The complex can also be split off by reversible change of the steric structure with substances (deforming eluents) that disrupt the hydrogen bonds or by hydrophobic interactions with, for example, urea, guanidine hydrochloride, detergents, or chaotropic ions. Here also high concentrations of urea or guanidine hydrochloride can be used for immunoadsorbents, frequently in combination with low pH. If using relatively drastic elution conditions, care should be taken that the conditions are changed immediately after elution by well-known methods.

A very highly specific method is the affinity elution. Its basic principle is to displace the matrix-bound ligand from the complex equilibrium by the presence of the free ligand in the mobile phase (elution buffer) and thus to elute the adsorbed complex partner from the column. For this purpose one may use substrates, effectors, inhibitors, cofactors, and other mostly low-molecular-weight ligands or their structural analogs. This method is largely applied for the group-specific ligands, for example the elution of cofactor dependent enzymes from 5'-AMP- or 2',5' ADP-supports by free NADH or NADPH or glycoproteins from lectin supports by free sugar.

The concentrations necessary for the elution are very low, often less than 10 mM. These low concentrations, as well as a relatively high ionic strength in a nonspecific elution with salts, are unequivocal evidence for highly specific complex formation at the matrix. Affinity elution can also be carried out by a gradient.

Pulse elution is better because the specific ligands are very expensive. With pulse elution, the ligand is dissolved at relatively high concentration (1-10 mM) in a small volume of the elution buffer, applied to the column, and subsequently further eluted with the buffer without ligand. The substance to be purified is eluted as intact ligand complex and must subsequently be purified from the ligand by well-known procedures.

A special method to elute a biomacromolecule is electrophoresis. It is especially suited for the desorption of antibodies and hormone binding proteins (Morgan et al., 1978b). A special apparatus is needed for this method (Dean et al., 1979). The method may also serve as an alternative for the elution of serum albumin from Cibacron Blue F3G-A supports (Morgan et al., 1978a).

Another method is the selective cleavage of the bond between the matrix and the ligand after adsorption of the substance to be purified. With this method, the ligand complex is eluted analogously to affinity elution. The method is used mainly for high-affinity complexes, where normal elution leads to denaturation. In this way, Cuatrecasas (1970) isolated the serum estradiol-binding proteins from estradiol-agarose by reductive cleavage of the azo-linkage with dithionite. A prerequisite for this method is a relatively easy cleavage of the support-ligand bond.

In some cases a specific effect of the borate buffer was also found. It is assumed that the borate ion interacts specifically with the ligand, such as sugar residues, and thus decreases the binding of the enzyme (Steers et al., 1971; Barker et al., 1972).

Determination of Immobilized Ligands

A part of the characterization of a support for affinity chromatography is to determine the concentration of immobilized ligands. The influence of ligand concentration was dealt with in more detail in a previous section. The indirect method is very simple and versatile; determination from the difference between the amount of ligand added to the coupling mixture and the amount found after the coupling procedure and the wash process in the washing solution.

Direct spectroscopic determination at the support is frequently used, provided that the gel suspension is relatively transparent, the ligand adsorbs at wavelengths above 260 nm, and a high-performance double-beam spectrophotometer is available. Comparing this method with others, Koelsch and coworkers (1975) have shown that the spectrophotometric and fluorimetric determination of bound proteins is characterized by simplicity and high reproducibility. Schurz and Rüdiger (1982) used for the determination of immobilized proteins the first derivative of the absorption spectrum. The signal heights in the absorption range are related to protein density in a proportional way.

Non-cross-linked agarose gels are relatively easy to solubilize (Failla and Santi, 1973). Subsequently, ligand concentration can be determined spectrophotometrically. Most methods first split off the ligand from the support by alkaline, acidic, or enzymatic hydrolysis, as well as by oxidative or reductive procedures. Subsequently the concentration of the ligand is determined by classic methods. A detailed review has been published by Turková (1978).

Capacity of an Affine Support

Not all matrix-bound ligands can form a complex with the complementary partner. Steric factors play a predominant role, but such factors as equilibrium time, flow rate, and methods of adsorption must be considered, too.

The capacity can be determined in both the batch and the column procedure. In the first case, a defined quantity (weight or volume) of the adsorbent m is incubated with a defined quantity of the affine partner with the concentration C_0, and after establishing the equilibrium the concentration of the nonbound proportion C is determined. The capacity is then calculated from Equation (13)

$$\text{Capacity} = \frac{C_0 - C}{m} \tag{13}$$

The determination is better and simpler with a column bed by frontal chromatography. Figure 6.13 gives a theoretical scheme for determination of the capacity by frontal analysis.

With this method the complementary partner is chromatographed continually with a defined concentration (C_0) via a small matrix quantity until saturation. The retention volume V_e, the volume at which the concentration of the complementary partner in the eluate increases drastically over a small volume, consists of the void volume V_o and the volume V from which the partner was removed either completely or partially through adsorption to the matrix. This method is the best way of determining an adsorption isotherm in which the capacity is determined with different C_0.

Regeneration and Storage of the Affine Support

Regeneration of the support material is a prerequisite to repeated use. The conditions of regeneration are dependent on the nature of the

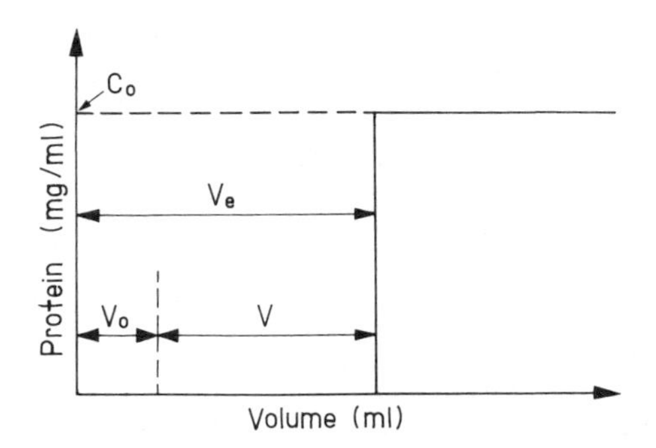

FIGURE 6.13 Theoretical scheme for the determination of the capacity of an affinity support by frontal analysis: C_0 = starting concentration, V_e = retention volume, V_o = void volume, V = volume of solution from which the substance was removed by the adsorbent.

affine gel and of the sample and must not change the chromatographic behavior. Sufficient washing with the elution buffer and subsequent equilibration with the starting buffer will frequently suffice. Often high concentrations, up to 6 M guanidine hydrochloride or ureas, are used for regeneration. If no specific methods are necessary, then a washing process alternating with alkaline (pH 8-9) and acidic (pH 4-5) buffer containing 0.5 M NaCl will be sufficient.

Since in many cases both the ligands and the matrix material are biopolymers, the affinity gels should be stored under mild conditions. To preserve its properties, the best way is to store the matrix at low temperatures in the presence of a bacteriostatic agent, such as 0.02% sodium azide, and with a minimum of supernatant.

REFERENCES

Amicon Corporation (1980). *Dye-Ligand Chromatography*. Amicon Corporation, Lexington, Massachusetts.

Barker, R., Olsen, K. W., Shaper, J. H., and Hill, R. L. (1972). Agarose derivatives of uridine diphosphate and N-acetylglucosamine for the purification of a galactosyltransferase. *J. Biol. Chem. 247*: 7135-7147.

Berliner, L. J. (ed.). (1979). *Spin-Labeling: Theory and Applications*, Vol. 2. Academic Press, New York.

Birkenmeier, G. (1981). Untersuchungen über die Wechselwirkung von Serumproteinen und trägerfixierten Cibacronblau F3G-A und anderen Triazinfarbstoffen. *Ph.D Thesis*, Section of Medicine, Karl-Marx-University, Leipzig.

Chance, B. (1943). The kinetics of the enzyme-substrate compound of peroxidase. *J. Biol. Chem. 151*: 553-562.

Cuatrecasas, P. (1970). Protein purification by affinity chromatography. Derivatizations of agarose and polyacrylamide beads. *J. Biol. Chem. 245*: 3059-3065.

Cuatrecasas, P. (1972). Affinity chromatography on macromolecules. *Advan. Enzymol. 36*: 29-89.

Cuatrecasas, P., Wilchek, M., and Anfinsen, C. B. (1968). Selective enzyme purification by affinity chromatography. *Proc. Nat. Acad. Sci. U. S. 61*: 636-643.

Dean, P. D. G., Quadri, F., Jessup, W., Bouriotis, V., Angal, S., Potuzak, H., Leatherbarrow, R. J., Miron, T., George, E., and Morgan, M. R. A. (1979). Design faults in affinity chromatography. In *Affinity Chromatography and Molecular Interactions*, J.-M. Egly (ed.). Editions-INSERM, Paris, pp. 321-344.

Ebert, B., Schwarz, D., and Lassmann, G. (1981). Study of Brownian rotation motion in dense solutions of hemoglobin. *Stud. Biophys. 82*: 105-112.

Ebert, B., Elmgren, H., and Hanke, T. (1982). ESR Investigation on the state of water in heterogeneous systems. *Stud. Biophys. 91*: 19-22.

Erickson, J., and Paucker, K. (1979). Purification of ethanol extracted human lymphoid interferons by Blue Sepharose chromatography. *Anal. Biochem. 98*: 214-218.

Failla, D., and Santi, D. V. (1973). A simple method for quantitating ligands covalently bound to agarose beads. *Anal. Biochem. 52*: 363-368.

Franks, F. (1975). The hydrophobic interaction. In *Water, a Comprehensive Treatise*, Vol. 4, F. Franks (ed.). Plenum Press, New York, pp. 1-94.

Frieden, E. (1975). Non-covalent interactions; key to biological flexibility and specificity. *J. Chem. Educ. 52*: 754-761.

Harvey, M. J., Lowe, C. R., Craven, D. B., and Dean, P. D. G. (1974). Affinity chromatography on immobilised adenosine 5'-monophosphate. 2. Some parameters relating to the selection and concentration of the immobilised ligand. *Eur. J. Biochem. 41*: 335-340.

Herrling, T., Klimes, N., Karthe, W., Ewert, U., and Ebert, B. (1982). EPR Zeugmatography with modulated magnetic field gradient. *J. Magn. Reson. 49*: 203-211.

Hethcote, H. W., and DeLisi, C. (1982). Determination of equilibrium and rate constants by affinity chromatography. *J. Chromatogr. 248*: 183-202.

Höhne, E., and Kretschmer, R.-G. (1982). New interpretation of helical structures in polypeptides. *Stud. Biophys. 87*: 23-28.

Höhne, E., and Kretschmer, R.-G. (1983). New interpertation and uniform description of secondary structures in proteins. *Stud. Biophys. 98*: 85-94.

Holroyde, M. J., Chesher, J. M. E., Trayer, J. P., and Walker, D. G. (1976). Studies on the use of Sepharose-N-(6-aminohexanoyl)-2-amino-2-deoxy-D-glycopyranose for the large-scale purification of hepatic glucokinase. *Biochem. J. 153*: 351-361.

Israelachvili, J., and Pashley, R. (1982). The hydrophobic interaction is long range, decaying exponentially with distance. *Nature 300*: 341-342.

Jeffrey, G. A. (1982). Hydrogen-bonding in amino acids and carbohydrates. In *Molecular Structure and Biological Activity*. J. F. Griffin and W. L. Dirax (eds.). Elsevier Scientific, Amsterdam.

Jost, R., Miron, T., and Wilchek, M. (1974). The mode of adsorption of proteins to aliphatic and aromatic amines coupled to cyanogen bromide-activated agarose. *Biochim. Biophys. Acta 362*: 75-82.

Karthe, W., and Wehrsdorfer, E. (1979). The measurement of inhomogeneous distributions of paramagnetic centers by means of EPR. *J. Magn. Reson. 33*: 107-111.

Kertesz, J. C., and Wolf, W. (1973). The operational characteristics of a stopped-flow system for ESR studies. *J. Phys. Educ. 6*: 1009-1014.

Klimes, N., Lassmann, G., and Ebert, B. (1980). Time-resolved EPR spectroscopy. Stopped flow EPR apparatus for biological application. *J. Magn. Reson. 37*: 53-59.

Koelsch, R., Lasch, J., Marquardt, J., and Hanson, H. (1975). Application of spectrophotometric methods to the determination of protein bound to agarose beads. *Anal. Biochem. 66*: 556-567.

Kuznetsov, A. N., and Ebert, B. (1974). Determination of the character of rotational motion of nitroxyl radicals in their slow rotational region. *Stud. Biophys. 44*: 173-182.

Lassmann, G., Ebert, B., Kuznetsov, A. N., and Damerau, W. (1973). Characterization of hydrophobic regions in proteins by spin labeling technique. *Biochim. Biophys. Acta 310*: 298-304.

Lassmann, G., Ebert, B., Damerau, W., Sklenar, G., Pommerening, K., and Buttgereit, K. (1974). Spin labeling studies of quaternary structure of matrix bound oxyhemoglobin. *Stud. Biophys. 46*: 45-53.

Lowe, C. R., and Dean, P. D. G. (1971). Affinity gel filtration. A new method for the rapid determination of apparent molecular weights of enzymes. *FEBS Lett. 18*: 31-34.

Lowe, C. R., and Dean, P. D. G. (1974). *Affinity Chromatography*. John Wiley & Sons, London.

Lyklema, J. (1982). Molecular interactions in affinity chromatography. In *Affinity Chromatography and Related Techniques*. T. C. J. Gribnau, J. Visser, and R. J. R. Nivard (eds.). Elsevier Scientific, Amsterdam, pp. 11-27.

Martini, G., Ottaviani, M. F., and Romanelli, M. (1983). The dynamics of a nitroxide radical in water adsorbed on porous supports studied by ESR. *J. Colloid Interface Sci. 94*: 105-113.

Mohr, P., Hanke, T., Kuhn, W., and Ebert, B. (1983). Mechanism studies of enzymatically formed tolidine blue and determination of peroxidatic activities. *Biomed. Biochim. Acta 42*: 663-672.

Morgan, M. R. A., Brown, P. J., Leyland, M. J., and Dean, P. D. G. (1978a). Electrophoresis: A new preparative desorption technique in affinity chromatography (and immunoadsorption) *FEBS Lett. 87*: 239-243.

Morgan, M. R. A., Kerr, E. J., and Dean, P. D. G. (1978b). Electrophoretic desorption: Preparative elution of steroid specific antibodies from immunoadsorbents. *J. Steroid Biochem. 9*: 767-770.

O'Carra, P., Barry, S., and Griffin, T. (1973). Spacer-arms in affinity chromatography: The need for a more rigorous approach. *Biochem. Soc. Trans. 1*: 289-290.

Ohno, K. (1982). Application of ESR imaging to a continuous flow method for study on kinetics of short-lived radicals. *J. Magn. Reson. 49*: 56-62.

Pommerening, K., Blanck, J., Mauersberger, K., Behlke, J., Honeck, H., Smettan, G., Ristau, O., and Rein, H. (1973). Studies on the characterization of matrix-bound solubilized human hemoglobin. *Abh. Deut. Akad. Wiss. Berlin, Kl. Chem., Geol. Biol.* 179-186.

Pommerening, K., Mohr, P., and Zorn, H. (1978). Haptoglobin preparation by affinity chromatography. In *Chromatography of Synthetic and Biological Polymers*, Vol. 2, R. Epton (ed.). Ellis Horwood Publishers, Chichester, pp. 293-297.

Pommerening, K., Kühn, M., Jung, W., Buttgereit, K., Mohr, P., Štamberg, J., and Beneš, M. (1979). Affinity chromatography of haemoproteins. 1. Synthesis of various imidazole containing matrices and their interaction with haemoglobin. *Int. J. Biol. Macromol. 1*: 79-88.

Pommerening, K., Ristau, O., Kühn, M., Mohr, P., Štamberg, J., and Beneš, M. (1980). Investigation of interaction between methemoglobin and polymethacrylates containing imidazole or hydrophobic side chains by means of EPR spectroscopy. *J. Polym. Sci., Polym. Symp. 68*: 79-88.

Record, M. T., Anderson, C. F., and Lohman, T. M. (1978). Thermodynamic analysis of ion effects on the binding and conformational equilibria of proteins and nucleic acids: The role of ion association or release, screening, and ion effects on water activity. *Q. Rev. Biophys. II*: 103-178.

Schurz, H., and Rüdiger, H. (1982). A spectrophotometric determination of protein immobilized to affinity gels. *Anal. Biochem. 123*: 174-177.

Scouten, W. H. (1981). *Affinity Chromatography. Bioselective Adsorption on Inert Matrices*. John Wiley & Sons, New York.

Steers, E., Cuatrecasas, P., and Pollard, H. (1971). The purification of β-galactosidase from escherichia coli by affinity chromatography. *J. Biol. Chem. 246*: 196-200.

Turková, J. (1978). *Affinity Chromatography*. Elsevier Scientific, Amsterdam.

Turková, J., Hubálková, O., Kriváková, M., and Čoupek, J. (1973). Affinity chromatography on hydroxyalkyl methacrylate gels. I. Preparation of immobilized chymotrypsin and its use in the isolation of proteolytic inhibitors. *Biochim. Biophys. Acta 322*: 1-9.

Turková, J., Bláha, K., and Adamová, K. (1982). Effect of concentration of immobilized inhibitor on the biospecific chromatography of pepsins. *J. Chromatogr. 236*: 375-383.

Van Oss, C. J., Absolom, D. R., and Neumann, A. W. (1980). Applications of net repulsive van der Waals forces between different particles, macromolecules or biological cells in liquids. *Colloids Surfaces 1*: 45-46.

Van Oss, C. J., Absolom, D. R., and Neumann, A. W. (1982). Role of attractive and repulsive van der Waals forces in affinity and hydrophobic chromatography. In *Affinity Chromatography and Related Techniques*, T. C. J. Gribnau, J. Visser, and R. J. F. Nivard (eds.). Elsevier Scientific, Amsterdam, pp. 29-37.

part III
BIOSPECIFIC AFFINITY CHROMATOGRAPHY (BIOADSORPTION)

7
Enzymes and Effectors

Biospecific affinity chromatography is based on reversible, more or less specific interactions between biologically active substances. In case of biomacromolecules this ability is due to well-defined sequences, areas, or pockets characterized by the chemical structure and steric arrangement of groups forming such regions. Therefore, only fit or adaptable molecules or parts of molecules are recognized, as schematically illustrated for enzyme-inhibitor interactions in Figure 7.1 by the examples of α-chymotrypsin and formyltryptophan, as well as carboxypeptidase A and glycyltyrosine. To isolate any biologically active substance (e.g., an enzyme or its effector), another able to interact with it is anchored covalently to the adsorbent particles by any of the methods described in Chapter 5. On this basis biospecific affinity chromatography has been developed in the last 20 years to an efficient method on the laboratory scale, and many monographs have been published dealing with this matter (Turková, 1974; Lowe and Dean, 1974; Epton, 1978; Hoffmann-Ostenhof et al., 1978; Egly, 1979; Turková, 1978; Gribnau et al., 1982). The field of application of biospecific affinity chromatography has come to include enzymology, immunology, and nucleic acid chemistry, as well as the separation of membrane ingredients (including receptors), viruses, and cells. A series of substance classes that can be isolated by this method or that have been used as ligands is shown in Table 5.1 of Chapter 5. The classification given here rests on the fact that the basis of biospecific affinity chromatography is the natural biochemical function of the different types of substances rather than their chemical structure. Such a schematic practice, however, does not exclude overlapping. Nevertheless, from a practical point of

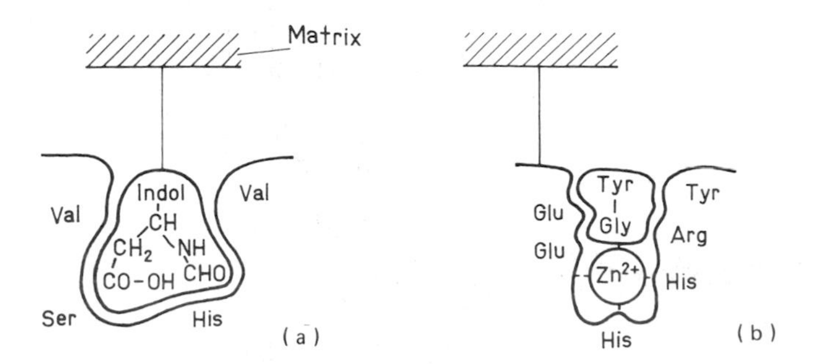

FIGURE 7.1 Schematic representation of biospecific complex formation. (a) Adsorption of α-chymotrypsin to matrix-bound formyltryptophan. (b) Adsorption of glycyltyrosine to matrix-bound carboxypeptidase A.

view in the following the biospecific affinity chromatography is discussed on this base.

POSSIBILITIES AND PROBLEMS

The natural ability of enzymes to recognize their substrates, effectors (activators and inhibitors), cofactors, or molecules with adequate groups has been used successfully in the last few years to separate these substances from biological materials of different origin, and an immense number of papers has been published in this field up to now. Figures 7.2b and 7.3b demonstrate, based on the principle of protease inhibitor interactions, the biospecific affinity chromatography of chymotrypsin by the lung trypsin inhibitor bound to Spheron 300 and, vice versa, the chromatography of the inhibitor by chymotrypsin bound to the same adsorbent material (Turková et al., 1973). In both cases the elution was done using pH gradients from 8.0 to 3.1. The chromatographic characteristics of the columns filled with unsubstituted Spheron 300 (Figs. 7.2a and 7.3a) points clearly to biospecific sorption as the basis of the separation efficiency as shown in Figures 7.2b and 7.3b.

In Table 7.1 some further examples are given that rest on enzyme inhibitor interactions that demonstrate the broad applicability of biospecific affinity chromatography in enzyme and inhibitor purification, respectively. For the general course and the efficiency of all these methods the same methodical demands are valid as for most other affinity chromatographic separation techniques described in later chapters. These are sufficient chemical and mechanical stability of the

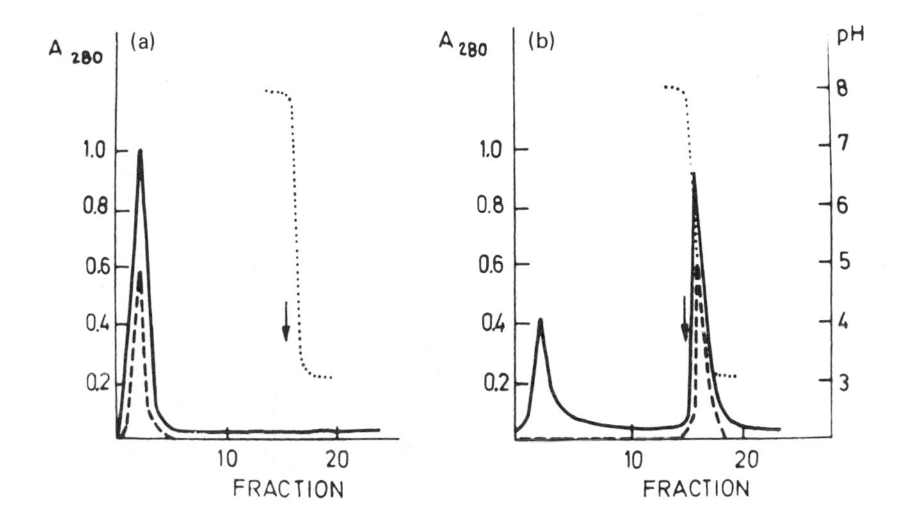

FIGURE 7.2 Chromatography of trypsin inhibitor on Spheron 300 (a) and chymotrypsin Spheron 300 (b) columns (10 X 1 cm). Trypsin inhibitor (30 mg) was applied to the columns and 6 ml fractions were collected at 10 min intervals: (—) adsorbance at 280 nm, (---) inhibitor activity, (···) pH. Vertical arrow, elution buffer changes from pH 8.1 (50 mM tris·HCl buffer) to pH 3.1 (approximately 100 mM acetic acid). (From Turková et al., 1973.)

adsorbent particles and good flow qualities of the column packings used.

The elution of adsorbed enzymes or effectors may be done by variation of milieu factors of the mobile phase (change of pH, salt concentration, and temperature or addition of detergents, organic solvents, or competing substances), as shown in Table 7.1. In all these cases conditions must be selected in such a manner that irreversible change of the matrix, of the ligand, or of the substance to be separated cannot occur.

AFFINITY CHROMATOGRAPHY AT SUBZERO TEMPERATURES

Enzyme purification might be carried out by affinity chromatography using real substrates as ligands if their conversion to the corresponding products could be stopped widely. In many cases this is possible

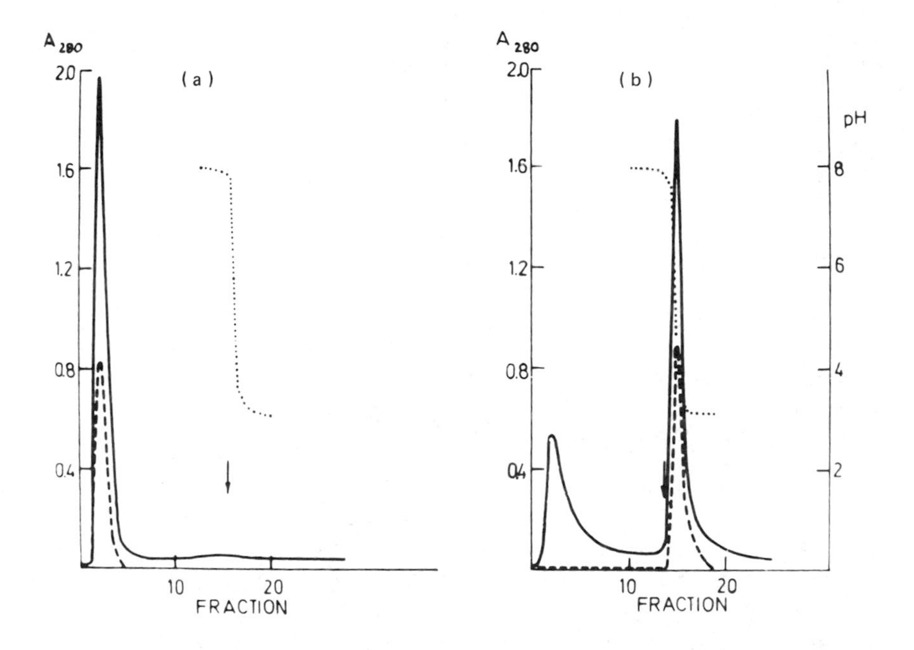

FIGURE 7.3 Chromatography of chymotrypsin on Spheron 300 (a) and trypsin inhibitor Spheron 300 (b) columns (10 X 1 cm). Chymotrypsin (30 mg) was applied to the columns and 6 ml fractions were collected at 10 min intervals: (—) adsorbance at 280 nm, (---) proteolytic activity, (···) pH. Vertical arrow, elution buffer changes from pH 8.0 (50 mM tris·HCl buffer) to pH 3.1 (approximately 100 mM acetic acid). (From Turková et al., 1973.)

at subzero temperatures at which the reaction rate of product formation can be sufficiently decreased. Such low-temperature procedures using, if possible, mixed solvents or highly salted aqueous solutions of low freezing point (<0°C) have been successfully applied in enzyme purification. As an example can serve the affinity binding of porcine pancreatic elastase toward Sepharose-bound L-trialanine-p-nitroanilide under nonturnover conditions (Balny et al., 1979; Balny and Douzou, 1979). This procedure was carried out at high salt concentrations at which the samples can be cooled to -14°C without any influence on enzyme substrate affinity.

For the elution of the enzyme, two alternatives are available: (a) the separation of the enzyme substrate interaction by simple heating. As a consequence the bound substrate ligand is converted into the product, (b) release at -14°C by using a precooled mixture of ethylene glycol and water (50% v/v) in case of porcine pancreatic elastase. For

TABLE 7.1 Examples of Enzymes and Inhibitors Separated on the Basis of Enzyme Inhibitor and Enzyme Substrate Interactions

Substance to be separated	Adsorbent	Elution	References
Adult culex chymotrypsin inhibitor	Larval culex chymotrypsin coupled to Sepharose	200 mM KCl/HCl, pH 2.0	Spiro-Kern and Chen (1978)
Bacterial pectate lyase	Cross-linked pectate	Salt gradient, pH 5.7	Visser et al. (1979).
Artemia salina α,α-trehalase	Lactose bound to amino-hexyl Sepharose 4B	500 mM NaCl, pH 6.0, 2°C	Bergami and Cacace (1978)
Glucokinase	Sepharose-N-(6-amino-hexanoyl)-2-amino-2-deoxy-D-glucopyranose	500 mM glucose, 20 mM tris·HCl, 10 mM KCl, 4 mM EDTA, 7.5 mM $MgCl_2$, 1 mM dithio-threitol, 5% (v/v) glycerol, pH 7.0	Trayer et al. (1978)
N^2-guanine RNA methyl-transferase	5-Adenosylhomocysteine Sepharose	NaCl gradient (0-500 mM), pH 8.5	Izzo and Gantt (1977)
Phenylalanine ammonia lyase	Phenylalanine/spacer/ Sepharose 4B	50 mM bicarbonate buffer, pH 9.3	Jack (1978)
Porcine pancreatic elastase	Sepharose-bound L-triala-nine p-nitroanilide	100 tris buffer con-taining 50 (v/v) ethylene glycol, pH 9.2, t = 14°C	Balny and Douzou (1979)

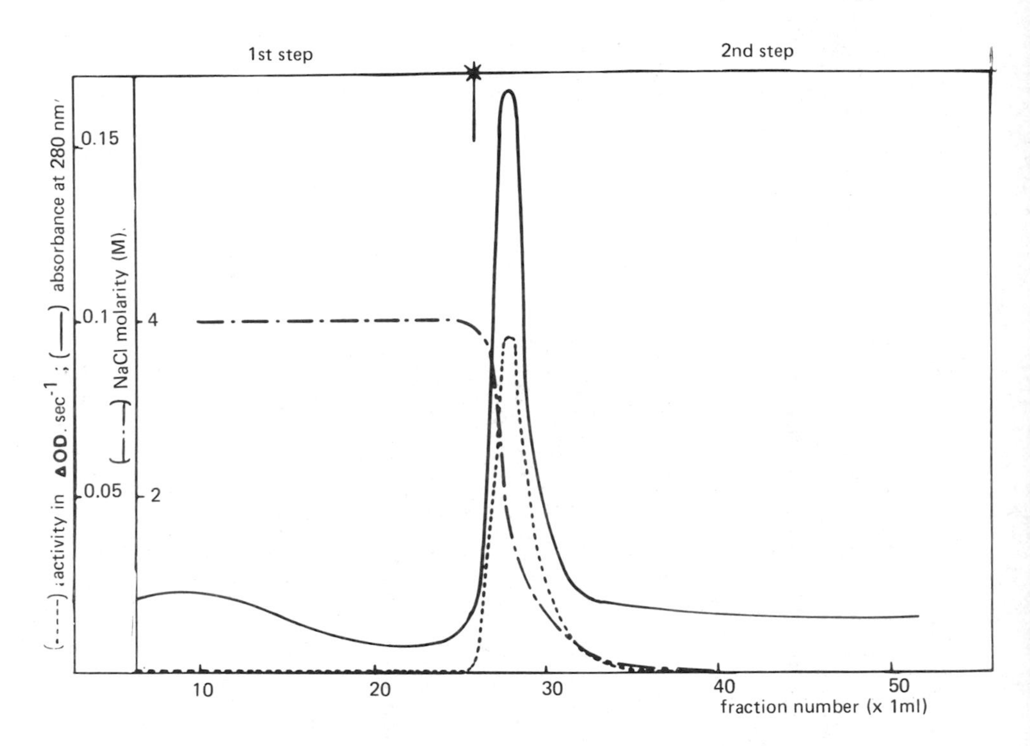

FIGURE 7.4 Elution pattern of elastase (0.2 mg applied) from the affinity column (2 X 1 cm) at -14°C. First step, 100 mM tris buffer containing 4 M NaCl, pH 9.2; second step, 100 mM tris buffer containing 50% (v/v) ethylene glycol, pH 9.2: (—) adsorbance at 280 nm; (---) activity in ΔA sec^{-1} measured at 410 nm (20°C) using succinyl-L-trialanine p-nitroanilide as substrate; (—·—) molarity of the fractions. (From Balny and Douzou, 1979.)

this variant an elution diagram demonstrating the general run is shown in Figure 7.4.

Because of the great methodological expense of such subzero procedures, it can be assumed that in the future their application will be limited to a few special cases in which separation cannot be carried out by another procedure at normal temperatures.

REFERENCES

Balny, C., and Douzou, P. (1979). Affinity chromatography at subzero temperatures, preliminary results, perspectives and problems. In *Affinity Chromatography and Molecular Interactions*, J.-M. Egly (ed.). Editions INSERM, Paris, pp. 99-108.

Balny, C., LeDoucen, C., Douzou, P., and Bieth, J. G. (1979). Affinity chromatography at sub-zero temperatures. A model study with porcine pancreatic elastase. *J. Chromatogr. 168*: 133-138.

Bergami, M., and Cacace, M. G. (1978). Direct coupling of reducing oligosaccharides to aminohexyl Sepharose: Purification of α,α-trehalase from Artemia salina. In *Affinity Chromatography*, O. Hoffmann-Ostenhof, M. Breitenbach, F. Koller, D. Kraft, and O. Scheiner (eds.). Pergamon Press, Oxford, pp. 129-132.

Egly, J.-M. (ed.) (1979). *Affinity Chromatography and Molecular Interactions*, Editions INSERM, Paris.

Epton, R. (ed.) (1978). *Chromatography of Synthetic and Biological Polymers*, Vol. 2, *Hydrophobic, Ion-Exchange and Affinity Methods*. Ellis Horwood Ltd., Chichester.

Gribnau, T. C. J., Visser, J., and Nivard, R. J. F. (eds.) (1982). *Affinity Chromatography and Related Techniques*, Elsevier Scientific, Amsterdam.

Hoffmann-Ostenhof, O., Breitenbach, M., Koller, F., Kraft, D., and Scheiner, O. (eds.) (1978). *Affinity Chromatography*. Pergamon Press, Oxford.

Izzo, P., and Gantt, R. (1977). Partial purification and characterization of an N^2-guanine RNA methyltransferase from chicken. *Biochemistry 16*: 3576-3581.

Jack, G. W. (1978). The affinity chromatography of phenylalanine degrading enzymes. In *Chromatography of Synthetic and Biological Polymers*, Vol. 2, R. Epton (ed.). Ellis Horwood Ltd., Chichester, pp. 277-281.

Lowe, C. R., and Dean, P. D. G. (1974). *Affinity Chromatography*. John Wiley & Sons, London.

Spiro Kern, A., and Chen, P. S. (1978). A chymotrypsin inhibitor from adult culex pipiens isolated by affinity chromatography. In *Affinity Chromatography*, O. Hoffmann-Ostenhof, M. Breitenbach, F. Koller, D. Kraft, and O. Scheiner (eds.). Pergamon Press, Oxford, pp. 119-122.

Trayer, I. P., Holroyde, M. J., Small, D. A. P., Trayer, H. R., and Wright, C. L. (1978). Design of affinity chromatography systems from free solution kinetics. In *Chromatography of Synthetic and Biological Polymers*, Vol. 2, R. Epton (ed.). Ellis Horwood Ltd., Chichester, pp. 159-178.

Turková, J. (1974). Affinity chromatography. *J. Chromatogr. 91*: 267-291.

Turková, J. (1978). *Affinity Chromatography*. Elsevier Scientific, Amsterdam.

Turková, J., Hubalková, O., Křiváková, M., and Čoupek, J. (1973). Affinity chromatography on hydroxymethacrylate gels. I. Preparation of immobilized chymotrypsin and its use in the isolation of proteolytic inhibitors. *Biochim. Biophys. Acta 322*: 1-10.

Visser, J., Maeyer, R., Topp, R., and Rombouts, F. (1979). Purification of pectate lyases on crosslinked pectate. In *Affinity Chromatography and Molecular Interactions*, J.-M. Egly (ed.). Editions INSERM, Paris, pp. 51-62.

8
Immunoaffinity Chromatography and Immunoassays

IMMUNOAFFINITY CHROMATOGRAPHY

Matrix bound antigens or antibodies are increasingly used as tools to separate or isolate complementary immunosubstances. This possibility rests on the fact that a given antigen (or biopolymer-bound hapten) is able to form complexes exclusively with antibodies that have been induced by themselves in the respective immunization procedure (Livingstone, 1974). Since immunocomponents (including monoclonal antibodies) can be highly purified by this method, nowadays immunoaffinity chromatography is an essential prerequisite for the development of new types of immunoassays as efficient analytical test methods.

The principle of antigen antibody interaction is illustrated in Figure 8.1. The dimeric antibody (immunoglobulin G is shown as example) is divided into so-called Fc fragments or constant regions and the antigen binding F(ab) fragments termed "variable regions." The antigen or hapten conjugate binding areas are arranged on the top of the variable regions as has been shown by means of x-ray diffraction studies (Davies et al., 1975). It is an essential feature that these areas vary from antibody to antibody in their amino acid topography, and consequently, only the complementary antigens can form real immunocomplexes. The forces that make immunocomplexes stick together are always in such cases a medley of noncovalent interactions, such as coulombic and van der Waals forces, salt and hydrogen bridges, and hydrophobic and charge transfer interactions. With regard to these characteristics, antigen-antibody complexes

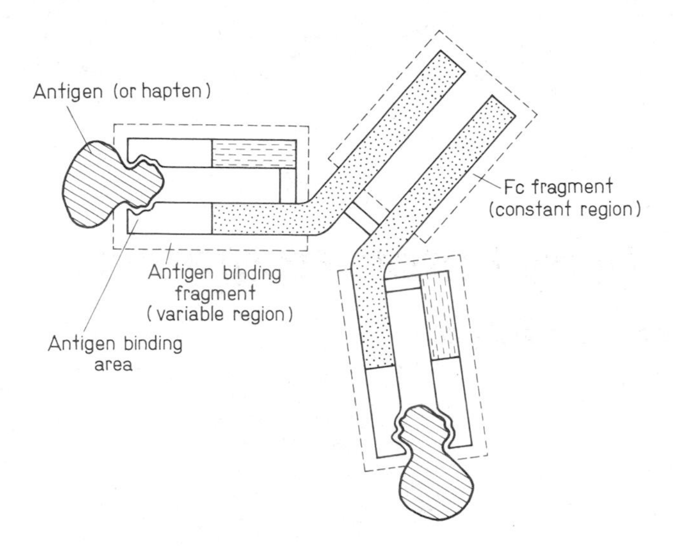

FIGURE 8.1 Antigen-antibody interaction.

are widely comparable with enzyme effector and enzyme substrate complexes (see Chap. 7).

For the practical application of immunosorption techniques, three fields may be distinguished (Kristiansen, 1978):

1. Removal of unwanted antigens from sample solutions using the batch method
2. Enrichment or isolation of antigen and antibody populations in relatively abundant supply using batch or column procedures
3. High recoveries of the fully active, precious antigens and antibodies on adsorbents of high stability that can be regenerated many times without loss of their activity

Moreover, immunoaffinity chromatography represents a sensitive tool for detection of immunoactive substances.

Strategies of the immunoaffinity chromatography for individual systems still have to be largely empirical since parameters of great importance, such as the extent of multipoint binding of the ligands under various coupling conditions, the true pH value, the charge distribution near the matrix backbone, and binding site of a coupled ligand within a given immunoadsorbent system, are not available at present (Kristiansen, 1978).

To separate any immunoactive substance by immunoaffinity chromatography, first the complementary substance is coupled to the matrix either with or without application of a spacer by means of any common method, which has to be selected in each case separately (see Chap. 5).

As matrix material have been used cellulose (Campbell and Weliky, 1967), polyacrylamide (Ternynck and Avrameas, 1972), and polyhydroxyethylmethacrylate (Tlaskalová et al., 1975); most frequently, however, agarose gels activated by CNBr are used. Good results were also obtained with composite gels, such as glutaraldehyde-activated agarose-polyacrylamide adsorbents for the isolation of rabbit, sheep, and bovine serum albumin antibodies, as well as of human and sheep gammaglobulin fraction II (Guesdon and Avrameas, 1976).

Interactions between multivalent antigen and a polyvalent antibody (e.g., bivalent IgG or multivalent IgM collected late during the immunoresponse) may be characterized by relatively high intrinsic affinity constants of 10^{10}-10^{11} M^{-1} referring to very stable complexes. This strong affinity usually demands the use of elution media causing destruction of biomacromolecules in many cases. It is characteristic of immunoaffinity chromatography that, generally, applicable desorption methods are still lacking. For this reason, desorption conditions that do not lead to irreversible deformation of any immunocomponent included in the chromatographic procedure must be found and optimized separately for each case. The most essential methods known may be subdivided in specific and nonspecific elution methods. A further possibility to decrease the complex stability consists in the preceding chemical modification of the immunoligand (Table 8.1).

When the matrix-bound ligand is a hapten, a hapten-analogous substance can be used as eluent in some cases. An example for such a specific desorption is the elution of pneumococcal polysaccharide antibodies from agarose based adsorbent by cellobiose (Cheng et al., 1970). Nonspecific desorption is possible by pH variation, chaotropic ions, urea, or guanidine hydrochloride. For example, the desorption of human haptoglobin adsorbed to Sepharose-bound human methemoglobin by 3 M guanidine hydrochloride solution is represented in the elution diagram of Figure 8.2. Although the isoantigen haptoglobin is not a real antigen, this model clearly illustrates the run of a simple immunoaffinity chromatographic elution procedure.

An example of a more complicated technique in immunoaffinity chromatography that demonstrates, moreover, the multiple possibilities of this method is the isolation of biologically active substances mediated by hapten conjugates (Wilchek and Gorecki, 1973; Wilchek, 1979). For isolation of trypsin the soybean trypsin inhibitor (STI) was dinitrophenylated (DNP-STI) by dinitrobenzene sulfonate and combined with trypsin. Then this complex was adsorbed to an antidinitrophenolate column and eluted under conditions (a) that break the antigen anti-DNP-antibody complex (Fig. 8.3) (Wilchek, 1979), or (b) that dissociate the trypsin directly from its inhibitor (Wilchek and Gorecki,

TABLE 8.1 Splitting of Antigen-Antibody Complexes Formed in Immunoaffinity Chromatographic Procedures

Method	Examples	References
Variation of the elution conditions		
Specific desorption	Elution of pneumococcal polysaccharide antibodies from agarose-based adsorbent with cellobiose	Cheng et al. (1973)
Nonspecific desorption	Isolation of immunoglobulins from mouse serum by means of protein A-Sepharose using a pH gradient	Ey et al. (1978)
Adjustment of pH	Isolation of collagen peptides at alkaline pH	Chidlow et al. (1974)
	Purification of human C3b inactivator by monoclonal-antibody affinity chromatography	Hsiung et al. (1982)
Using chaotropic ions (CCl_3COO^-, CF_3COO^-, SCN^-)	Desorption of anti-A antibody from matrix-bound partially deacetylated blood group substance A by 0-3 M CF_3COO^- gradient at pH 7.4	Kristiansen (1974)
By means of urea or guanidine hydrochloride	Separation of human haptoglobin from Sepharose-bound methemoglobin	Pommerening et al. (1978)
Chemical modification of the Original ligand structure		
Moderate chemical modification	Oxidation, alkylation, or nitration of amino acids of the matrix-bound immunoligand	Murphy et al. (1976)
Using immunoglobulin fragments as ligands	F(ab) fragments obtained by papain digestion of IgG	Fey (1975)

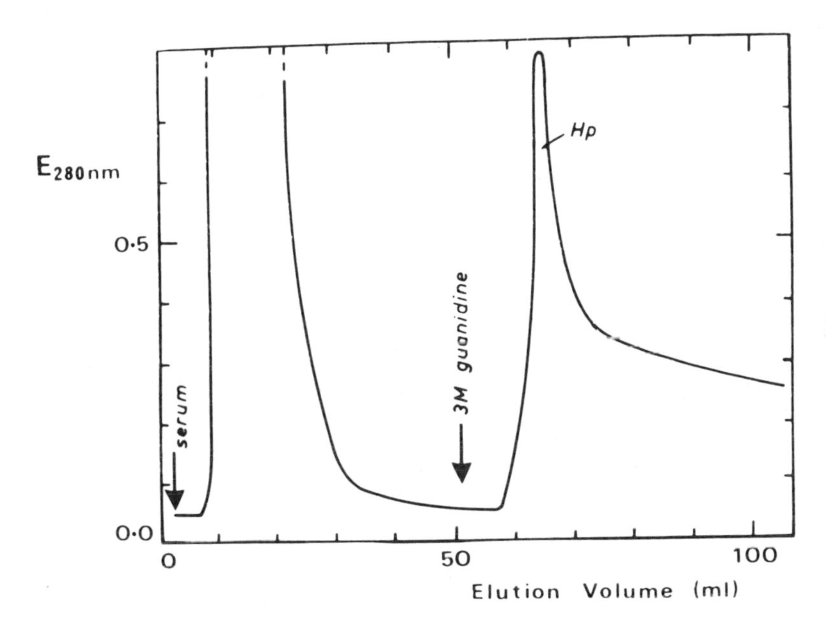

FIGURE 8.2 Haptoglobin isolation using a methemoglobin-Sepharose column containing 10 ml gel in 0.1 M phosphate buffer, pH 7.0. The bound haptoglobin was eluted with 3 M guanidine-HCl. (From Pommerening et al., 1978.)

1973). When (a) is applied, the eluted yellow hapten conjugate-trypsin complex can be separated into its components by bringing it to pH 2 and then passing it through a Sephadex G75 column. This method using a hapten conjugate as mediator has also been applied with success to the separation of peptides and insulin receptor (Wilchek, 1979).

IMMUNOASSAYS

The immunoassay is a highly sensitive analytic method suitable for the identification and quantitative determination of various substances, such as drugs and their metabolites, hormones, virus-specific immunoglobulins, and different proteins, including enzymes. The heterogenic immunoassay (the so-called ELISA technique; enzyme-linked immunosorbent assay) is mentioned briefly here because the basic principle rests on recognition phenomena and corresponds in this regard to the first step (adsorption step) of affinity chromatographic procedures. The function of such test systems may be demonstrated by the so-called

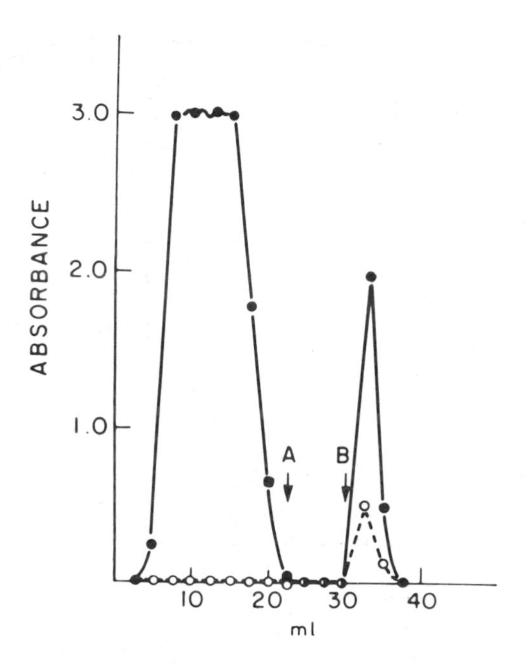

FIGURE 8.3 Isolation of DNP-STI-trypsin complex on anti-DNP-Sepharose. A mixture of 3 mg DNP-STI and 10 mg of trypsin in 1 ml of 0.1 M NH_4HCO_3 was applied to 0.5 X 10 cm column of anti-DNP-Sepharose. After washing with 0.1 M NH_4HCO_3 and water (a) elution was started with 10% formic acid (b). Adsorbance at 280 nm (●-●-●); adsorbance at 335 nm (o-o-o); DNP = dinitrophenyl; STI = soybean trypsin inhibitor. (From Wilchek, 1979.)

competitive immunoassays as proposed first by Yalow and Berson (1959). In Figure 8.4, for example, the determination of thyroxine (T_4) in human serum is represented schematically (Kleinhammer et al., 1978, 1979; Borner, 1979). In the first step (immunosorption) the sample solution and a known quantity of thyroxine labeled with the covalently bound enzyme peroxidase are pipetted in an antibody-coated plastic tube (e.g., polypropylene or polysterene). The sample antigen and the labeled antigen compete with the solid-phase-bound antibodies, forming antigen-antibody complexes. After removing the unbound antigen molecules by washing (separation step), a chromogen [2,2-di-(3-ethylbenzthiozoline sulfonate-6)] and H_2O_2 are added. The peroxidase label of the bound antigen then catalyzes dye formation, which can be followed up spectrophotometrically at

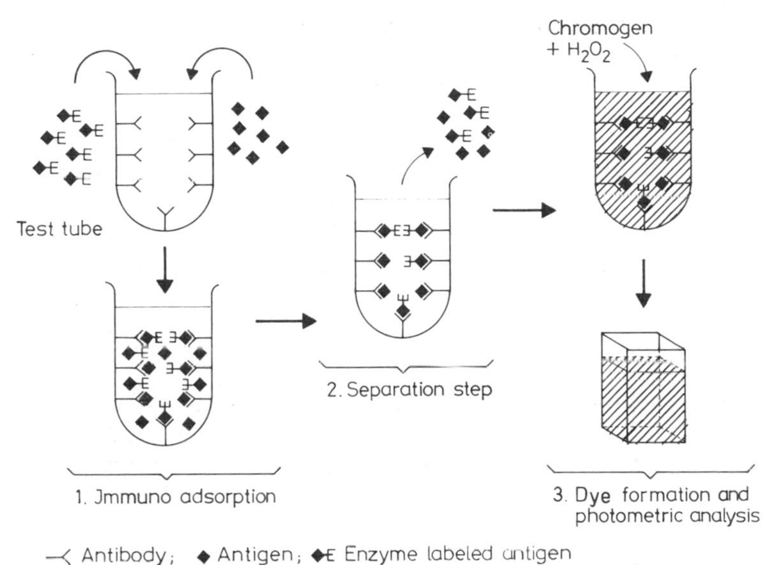

FIGURE 8.4 Schematic representation of the determination of thyroxin by the ELISA technique.

405 nm in the case of this chromogen. The extinction determined after a given time correlates inversely with thyroxine concentration in the sample solution. The exact value is found by means of a standard curve that must be ascertained before the beginning of measurements using a set of solutions containing the antigen and labeled antigen in known concentrations.

Besides enzymes as marker substances (e.g., peroxidase, glucose oxidase, alkaline phosphatase, and catalase), other types are described in the literature and applied in practice as sensitive analytic resources for immunoassays (Schall and Tenoso, 1981). A selection of usual markers that are principally anchored covalently with the antigen or antibody is given in Table 8.2. More detailed information is provided in the literature (Vogt, 1978, 1979).

The highly developed marker techniques are analytic tools also of general interest for research work in the field of affinity chromatography because they provide excellent possibilities for fundamental investigations on a molecular level. In such a manner the mobility and movement of an affine component, structural differences of antigens, antibodies, and other proteins in free and bound states, as well as the kinetics of complex formation, can be followed step by step very

TABLE 8.2 Selected Immunoassays and Markers

Marker	Type of immunoassay	Method of detection	References
Radioactive isotopes (e.g., ^{3}H and ^{125}I)	Radio	Scintillation spectrometry; radiation counting	Yalow and Berson (1959)
Enzymes (e.g., peroxidase and glucose oxidase)	Enzyme	Spectrophotometry	Oellerich (1980)
Fluorochromes (e.g., fluorescein and nicotine amide derivatives)	Fluorescent	Fluorescent spectrophotometry, fluorescence polarization	Sedlacek et al. (1979); Soini and Hemmila (1979).
Free radicals (spin markers)	Spin	Electron spin resonance	Piette and Hsia (1979); Shdanov (1981)
Dye sol particles (commercial dyestuffs)	Dye	Spectrophotometry	Gribnau et al. (1982)

exactly during an affinity chromatographic procedure by means of adequate measurement methods. It must be assumed, therefore, that the label technique developed for immunoassays will also influence the progress of affinity chromatography in the future.

REFERENCES

Borner, K. (1979). Die Bestimmung des Gesamtthyroxins im Serum. Erfahrungen mit einem heterogenen Immunoassay (ELISA). In *Praktische Anwendung des Enzymimmunoassay in der Klinischen Chemie und Serologie*, W. Vogt (ed.). Georg Thieme Verlag, Stuttgart, pp. 20-30.

Campbell, D. H., and Weliky, N. (1967). Immunosorbents: Preparation and use of cellulose derivatives. In *Methods in Immunology and Immunochemistry*, Vol. 1, C. A. Williams and M. W. Chase (eds.). Academic Press, New York, pp. 378-381.

Cheng, W. C., Fraser, K. J., and Haber, E. (1973). Fractionation of antibodies to the pneunococcal polysaccharides by affinity chromatography. *J. Immunol. 111*: 1677-1689.

Chidlow, J. W., Bourne, A. J., and Bailey, A. J. (1974). Production of hyperimmuno serum against collagen and its use for the isolation of specific collagen peptides on immunosorbent columns. *FEBS Lett. 41*: 248-252.

Davies, D. R., Padlan, E. A., and Segal, D. M. (1975). Three-dimensional structure of immunoglobulins. *Annu. Rev. Biochem. 44*: 639-667.

Ey, P. L., Prowse, S. J., and Jenkin, C. R. (1978). Isolation of pure IgG_1, IgG_{2a}, and IgG_{2b} immunoglobulins from mouse serum using protein A-Sepharose. *Immunochemistry 15*: 429-436.

Fey, H. (1975). A simple procedure for the production of F(ab) from bovine IgG as an adsorbent in the preparation of class specific anti-immunoglobulin. *Immunochemistry 12*: 235-239.

Gribnau, T., Roeles, F., van de Biezen, J., Leuwering, J., and Schuurs, A. (1982). The application of colloidal dye particles as label in immunoassays: Disperse(d) dye immunoassay ("DIA"). In *Affinity Chromatography and Related Techniques*, T. C. J. Gribnau, J. Visser, and R. J. F. Nivard (eds.). Elsevier Scientific, Amsterdam, pp. 411-424.

Guesdon, J. L., and Avrameas, S. (1976). Polyacrylamide-agarose beads for the preparation of effective immunosorbents. *J. Immunol. Methods 11*: 129-133.

Hsiung, L., Barclay, A. N., Brandon, M. R., Sim, E., and Porter, R. R. (1982). Purification of human C3b inactivator by monoclonal antibody affinity chromatography. *Biochem. J. 20*: 293-298.

Kleinhammer, G., Lenz, H., Linke, R., and Staehler, F. (1978). Enzymimmunologische Bestimmung von T_4 und T_3 in Anti-körper beschichteten Gläschen. In *Enzymimmunoassay — Grundlagen und praktische Anwendung*, W. Vogt (ed.). Georg Thieme Verlag, Stuttgart, pp.. 42-51.

Kleinhammer, G., Deutsch, G., Linke, R., and Staehler, F. (1979). Enzymimmunologische Bestimmung des Thyroxin-Bindungsindexes in Antikorper beschichteten Gläschen. In *Praktische Anwendung des Enzymimmunoassays in der klinischen Chemie und Serologie*, W. Vogt (ed.). Georg Thieme Verlag, Stuttgart, pp. 43-49.

Kristiansen, T. (1974). Studies on blood group substances. V. Blood group substance A coupled to Sepharose as an immunosorbent. *Biochim. Biophys. Acta 362*: 567-574.

Kristiansen, T. (1978). Matrix bound antigens and antibodies. In *Affinity Chromatography*, O. Hoffmann-Ostenhof, M. Breitenbach, F. Koller, D. Kraft, and O. Scheiner (eds.). Pergamon Press, Oxford, pp. 191-206.

Livingstone, D. M. (1974). Immunoaffinity chromatography of proteins. *Methods Enzymol. 34*: 723-731.

Murphy, R. F., Imam, A., Hughes, A. E., McGucken, M. J., Buchanan, K. D., Conlon, J. M., and Elmore, D. T. (1976). Avoidance strongly chaotropic eluents for immunoaffinity chromatography by chemical modification of immobilized ligands. *Biochim. Biophys. Acta 420*: 87-96.

Oellerich, M. (1980). Enzyme immunoassays in clinical chemistry: Present status and trends. *J. Clin. Chem. Clin. Biochem. 18*: 197-208.

Piette, L. H., and Hsia, J. C. (1979). Spin labeling in biomedicine. In *Spin Labeling. II. Theory and Applications*, L. C. Berliner (ed.). Academic Press, New York, pp. 247-290.

Pommerening, K., Mohr, P., and Zorn, H. (1978). Haptoglobin preparation by affinity chromatography. In *Chromatography of Synthetic and Biological Polymers*, Vol. 2, R. Epton (ed.). Ellis Horwood Ltd., Chichester, pp. 293-297.

Schall, Jr., R. F., and Tenoso, H. J. (1981). Alternatives to radioimmunoassay: Labels and methods. *J. Clin. Chem. 27*: 1157-1164.

Sedlacek, H. H., Muck, K. F., Rehkopf, R., Baudner, S., and Seiler, F. R. (1979). A new method for fluorescence immunoassay using plane surface solid phases (FIAPS). *J. Immunol. Methods 26*: 11-24.

Shdanov, R. J. (1981). *Paramagnetic Models of Biological Active Compounds* (in Russian). Sci. Publ. Moscow, pp. 213-263.

Soini, E., and Hemmila, J. (1979). Fluoroimmunoassay: Present status and key problems. *Clin. Chem. 25*: 353-361.

Ternynck, T., and Avrameas, S. (1972). Polyacrylamide-protein immunoadsorbents prepared with glutaraldehyde. *FEBS Lett. 23*: 24-28.

Tlaskalová, H., Tučkova, L., Křivaková, M., Rejnek, J., and Čoupek, J. (1975). Receptor-specific fractionation of immuno-competent cells and purification of antibodies on hydroxyalkyl methacrylate immunosorbents. *Immunochemistry 12*: 801-805.

Vogt, W. (ed.) (1978). *Enzymimmunoassay — Grundlagen und praktische Anwendung*. Georg Thieme Verlag, Stuttgart.

Vogt, W. (ed.) (1979). *Praktische Anwendung des Enzymimmunoassay in der Klinischen Chemie und Serologie*. Georg Thieme Verlag, Stuttgart.

Wilchek, M. (1979). Antibody and avidin columns for the isolation of biologically active compounds. In *Affinity Chromatography and Molecular Interactions*, J.-M. Egly (ed.). Editions INSERM, Paris, pp. 187-196.

Wilchek, M., and Gorecki, M. (1973). A new approach for the isolation of biologically active compounds by affinity chromatography: Isolation of trypsin. *FEBS Lett. 31*: 149-152.

Yalow, R. S., and Berson, S. A. (1959). Assay of plasma insulin in human subjects by immunological methods. *Nature 184*: 1648-1649.

9
Nucleic Acids

The formation and action of nucleic acids under biological conditions involves defined mutual contacts as well as their specific interactions with several enzymes and other proteins. These recognition phenomena have been used in the last few years to develop efficient affinity chromatographic procedures for their separation and purification. Nucleic acids are enabled, moreover, to form reversible contacts of nonbiological origin, which has led to further affinity techniques (Table 9.1), such as the nucleic acid dye chromatography (Bünemann and Müller, 1978). In the following, some selected methods for nucleic acid separation and detection are discussed in detail, but their application for separation of enzymes and proteins have not been considered.

POLY(A)-AFFINITY CHROMATOGRAPHY

Poly(A) sequences ranging in size from less than 20 to 250 bases in length occur at the 3' end of mRNA molecules of widely divergent organisms, including mammals, birds, insects, higher plants, viruses, yeasts, bacteria, and mitochondria (Ron et al., 1976; Haff and Bogorad, 1976). Although the physiological function of these poly(A) fragments is still unknown, they have led to the development of methods for their purification and separation from other cellular RNAs by affinity chromatography. The underlying concept utilizes the ability of these poly(A) strands to form selective hybrids with matrix-bound synthetic oligo(dT) (Nakazako and Edmonds, 1972; Aviv and Leder, 1972)

TABLE 9.1 Affinity Techniques for Nucleic Acid Detection and Separation[a]

Method	Application	Matrix
Poly(A)-affinity chromatography	Isolation of mRNA	Oligo(dT)-cellulose
Affinity chromatography by poly(A)-agaroses	Poly(A)-binding RNA,	Poly(A)-agarose
cDNA-cellulose affinity chromatography (including paper technique)[b]	Specific detection and isolation of mRNA by hybridization	cDNA-cellulose
Affinity chromatography by lysine-agarose[c, d]	rRNA, double-stranded DNA	Lysine-agarose
Polysome immunoprecipitation[c, e]	Isolation of specific polysomes with their mRNAs	Specific antibodies
Mercury affinity chromatography	Covalent chromatography of mercury-substituted RNAs	Sulfhydryl supports
Nucleic acid dye chromatography (including its application for affinity partition and affinity electrophoresis)	Isolation and detection	Dye supports[f]

[a]For references see text; a review including most of the techniques was given by Taylor and Hamilton (1979).
[b]Alwine et al. (1977).
[c]Not discussed in the following.
[d]Jones et al. (1976).
[e]Taylor and Tse (1976).
[f]See Chapter 12.

and polyuridylate [poly(U)] (Lindberg and Persson, 1972; Adesnick et al., 1972). The usually applied matrices are oligo(dT) (strands 12-18 nucleotides long) covalently attached to microgranular cellulose and poly(U) (strands ∿100 nucleotides long) bound to agarose (Sepharose 4B). Both types show differences with regard to their chromatographic properties. Taylor and Tse (1976), for example, have found a higher recovery of purer material of rat liver albumin mRNA in their chromatographic investigations if a poly(U) matrix was used instead of an oligo(dT) matrix. A more general advantage of poly(U)-gel is also the better interaction with short terminal poly(A) strands (Shapiro and Schimke, 1975), whereas efficient binding to oligo(dT) requires poly(A) sequences at least 75-100 nucleotides long. Purification of mRNAs using poly(A)-affinity chromatography represents the first stage of a general method for cloning any eukaryotic structural gene sequence for which the specific mRNA contains poly(A) (Higuchi et al., 1976).

When oligo(dT)-cellulose is used as adsorbent material an RNA-solution containing a moderately high salt concentration (e.g., 0.1-0.5 M NaCl and 0.1-0.5% sodium dodecylsulfate buffered to a neutral pH) is usually passed slowly through the column. The bound RNA is then eluted with a low salt buffer by thermal melting or denaturing solvents and recovered by NaCl-ethanol precipitation (Bantle et al., 1976; Taylor and Hamilton, 1979). Contaminants of the eluate by other nucleic acids are removed by its repeated passing through the column.

For fractionation of poly(A)-containing RNAs on the basis of poly(A) length, thermal elution chromatography in buffer solutions containing intermediate salt concentrations have been employed (Astell et al., 1973; Bantle et al., 1976).

Poly(U)-agarose has in contrast to oligo(dT)-cellulose, various advantages that have been mentioned already. A further advantage is that nonspecific adsorption of nucleic acids is practically not observed. Analysis of mRNA purified on poly(U)-Sepharose 4B by sucrose gradient centrifugation and polyacrylamide electrophoresis showed, for example, no contamination by 28S rRNA and 1-2% maximum of 18S rRNA (Shapiro and Schimke, 1975).

Poly(U)-agaroses have been used advantageously in the last few years to separate and purify various poly(A)-containing RNAs. An interesting example is the single-step separation of plant nucleic acids into DNA, rRNA, sRNA, poly(A) RNAs, and oligonucleotides (Grotha, 1976).

POLY(A)-AGAROSES AS AFFINITY SUPPORTS

Poly(A)-agaroses are prepared by covalent coupling of polyadenylic acid chains to agarose. By the cyanogen bromide method the resulting multipoint covalent attachment via N^6-amino groups of the base residues is more stable than single-point attachment by terminal free phosphate groups. In the case of poly(A)-Sepharose 4B, the length of this chain amounts to approximately 100 nucleotides, which acts as its own spacer and ensures good binding capacity. Poly(A)-agaroses may serve as adsorbents to isolate poly(U) strands containing RNA (Bajszar et al., 1976; Burdon et al., 1977; Katinakis and Burdon, 1978) including viral RNA (Yogo and Wimmer, 1975).

DETECTION AND ISOLATION OF mRNA BY cDNA-CELLULOSE AFFINITY CHROMATOGRAPHY

Specific cellulose-bound cDNA adsorbents (DNA complementary to mRNA) have been applied for the detection and isolation of both ovalbumin and globin mRNA via biospecific hybridization (Venetianer and Leder, 1974; Wood and Lingrel, 1977; Smith et al., 1978; Ross, 1978; Rhoads and Hellmann, 1978). Analogously, globin cDNA-cellulose was used for the isolation of nuclear precursor molecules to the cytoplasmic globin mRNA (Smith et al., 1978; Ross, 1978). Further applications of this method are the preparation of tissue-specific and commonly shared mRNAs by cDNA-cellulose synthesized from total unfractionated liver or hepatoma mRNAs (Hirsch et al., 1978) and the partial purification of specific ovalbumin gene sequences by ovalbumin cDNA-cellulose (Anderson and Schimke, 1976).

NUCLEIC ACID DYE CHROMATOGRAPHY

Adsorbents on polyacrylamide bases for the separation of several DNA species have been developed by Bünemann and Müller (1978). By copolymerization certain base-specific dyes, such as the guanine-cytosine-specific phenyl neutral red, and adenine-thymine-specific malachite green are immobilized as ligands. It is assumed that the malachite green ligand contacts one of the two grooves of the DNA helix, as is assumed similarly for other adenine-thymidine-specific dyes (Müller et al., 1975). For this reason the malachite green adsorbent can be applied for separation of double-stranded nucleic acids from single-stranded or partially single-stranded nucleic acids. For example, Figure 9.1 shows the elution of three different bacterial double-stranded DNA adsorbed by a malachite green-substituted column.

The guanine-cytosine-specific phenyl neutral red binds to DNA by intercalation (Müller and Gautier, 1975). Its planar tricyclic chromophore intercalates preferentially between two neighboring guanidine-cytosine base pairs of double-stranded DNA. The formation of such sandwichlike complexes of planar dyes with helical nucleic acids is facilitated energetically for supercoiled DNA species (Bauer and Vinograd, 1968). Phenyl neutral red adsorbents, therefore, may be used successfully for the separation and isolation of circular DNA molecules of plasmids and viruses, which often form the supercoiled conformation. The

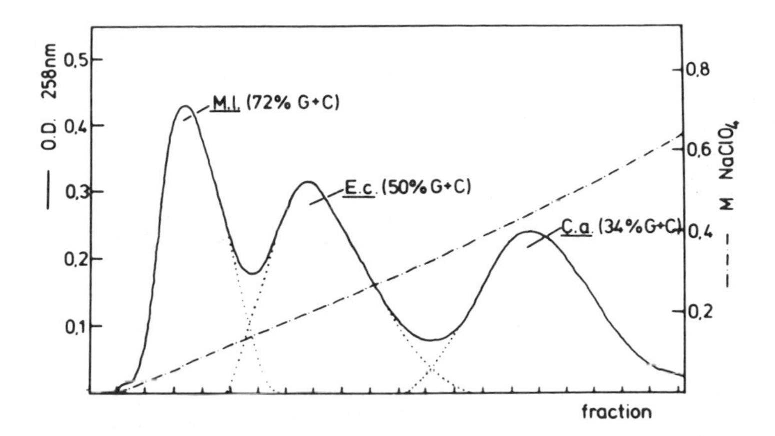

FIGURE 9.1 Elution of 1 mg of a mixture of three bacterial DNAs of an average molecular weight of about 700,000 from a malachite green-substituted column: 1.5 X 16 cm, 5.8 ml/hr flow rate at room temperature. The DNA mixture contained DNAs of *Clostridium acidiurici* (34% G + C), *Escherichia coli* (50% G + C), and *Micrococcus lutens* (72% G + C) in an optical density ratio of 1:1:1 at 258 nm. The column was equilibrated and operated with 10 mM sodium phosphate buffer, pH 6.0, 1 mM EDTA. Elution of DNA was performed by a gradient of sodium perchlorate in the same buffer; G = guanine, C = cytosine. (From Bünemann and Müller, 1978.)

principle of dye adsorption chromatography is also applicable for two-phase partition and gel electrophoretic procedures (Müller et al., 1982).

REFERENCES

Adesnik, M., Salditt, M., Thomas, W., and Darnell, J. E. (1972). Evidence that all messenger RNA molecules (except histone messenger RNA) contain poly(A) sequences and that the poly(A) has a nuclear function. *J. Mol. Biol. 71*: 21-30.

Alwine, J. C., Kemp, D. J., and Stark, G. R. (1977). Method for detection of specific RNAs in agarose gels by transfer to diazobenzyloxymethyl-paper and hybridization with DNA probes. *Proc. Nat. Acad. Sci. U.S. 74*: 5350-5354.

Anderson, J. N., and Schimke, R. T. (1976). Partial purification of ovalbumin gene. *Cell 7*: 331-338.

Astell, C. R., Doel, M. T., Jahnke, P. A., and Smith, M. (1973). Further studies on the properties of oligonucleotide cellulose columns. *Biochemistry 12*: 5068-5074.

Aviv, H., and Leder, P. (1972). Purification of biologically active globin messenger RNA by chromatography on oligothymidylic acid-cellulose. *Proc. Nat. Acad. Sci. U.S. 69*: 1408-1412.

Bajszar, G., Samanina, O. P., and Georgiev, G. P. (1976). On the nature of 5' termini in nuclear pre mRNA of Ehrlich carcinoma cells. *Cell 9*: 323-332.

Bantle, J. A., Maxwell, J. H., and Hahn, W. E. (1976). Specificity of oligo(dT)-cellulose chromatography in the isolation of polyadenylated RNA. *Anal. Biochem. 72*: 413-427.

Bauer, W., and Vinograd, J. (1968). The interaction of closed circular DNA with intercalative dyes. I. The superhelix density of SV 40 DNA in the presence and absence of dye. *J. Mol. Biol. 33*: 141-171.

Bünemann, H., and Müller, W. (1978). Synthesis of a novel material for affinity chromatography of biopolymers and its application for fractionation of nucleic acids. In *Affinity Chromatography*, O. Hoffmann-Ostenhof, M. Breitenbach, F. Koller, D. Kraft, and O. Scheiner (eds.). Pergamon Press, Oxford, pp. 353-356.

Burdon, R. H. Shenkin, A., Douglas, Y. T., and Smillie, E. J. (1977). Poly(A)-binding RNAs from nuclei and polysomes of BHK-21 cells. *Biochim. Biophys. Acta 474*: 254-267.

Grotha, R. (1976). Column chromatographic separation of plant total nucleic acids into DNA, rRNA, sRNA oligonucleotides and poly(A)-RNA in single step. *Biochem. Physiol. Pflanz. 170*: 273-277.

Haff, L. A., and Bogorad, L. (1976). Poly(adenylic acid)-containing RNA from plastids of maize. *Biochemistry 15*: 4110-4115.

Higuchi, R., Paddock, G. V., Wall, R., and Salser, W. (1976). A general method for cloning eukaryotic structural gene sequences. *Proc. Nat. Acad. Sci. U.S. 73*: 3146-3150.

Hirsch, F. W., Nall, K. N., Spohn, W. H., and Busch, H. (1978). Enrichment of special Novikoff hepatoma and regenerating liver mRNA by hybridization to cDNA-cellulose. *Proc. Nat. Acad. Sci. U.S. 75*: 1736-1739.

Jones, D. S., Lundgren, H. K., and Jay, F. T. (1976). The separation of ribonucleic acids from Escherichia coli on lysine-agarose. *Nucleic Acids Res. 3*: 1569-1576.

Katinakis, P. K., and Burdon, R. H. (1978). Non-polyadenylated messenger ribonucleic acids with affinity for polyadenylyl-Sepharose from Friend murine leukaemia cells. *Biochem. Soc. Trans. 6*: 757-758.

Lindberg, U., and Persson, T. (1972). Isolation of mRNA from KB-cells by affinity chromatography on polyuridylic acid covalently linked to Sepharose. *Eur. J. Biochem. 31*: 246-254.

Müller, W., and Gautier, F. (1975). Interactions of heteroaromatic compounds with nucleic acids. A·P-specific non-intercalating DNA-ligands. *Eur. J. Biochem. 54*: 385-394.

Müller, W., Bünemann, H., and Dattagupka, N. (1975). Interactions of heteroaromatic compounds with nucleic acids. 2. Influence of substitutents on the base and sequence specificity of intercalating ligands. *Eur. J. Biochem. 54*: 279-291.

Müller, W., Bünemann, H., Schuetz, H.-J., and Eigel, A. (1982). Nucleic acid interacting dyes suitable for affinity chromatography, partitioning and affinity electrophoresis. In *Affinity Chromatography and Related Techniques*, T. C. J. Gribnau, J. Visser, and R. J. F. Nivard (eds.). Elsevier Scientific, Amsterdam, pp. 437-444.

Nakazoka, H., and Edmonds, M. (1972). The isolation and purification of rapidly labeled polysome-bound ribonucleic acid on poly-thymidylate cellulose. *J. Biol. Chem. 253*: 3365-3367.

Rhoads, R. E., and Hellman, G. M. (1978). Chromatography of ovalbumin messenger ribonucleic acid on complementary deoxyribonucleic acid-cellulose. *J. Biol. Chem. 253*: 1687-1693.

Ron, A., Horovitz, O., and Sarov, J. (1976). Fractionation of RNA from tetrahymena by affinity chromatography on poly(U)-Sepharose. *J. Mol. Evol. 8*: 137-142.

Ross, J. (1978). Purification and structural properties of the precursors of the globin messenger RNAs. *J. Mol. Biol. 119*: 21-35.

Shapiro, D. J., and Schimke, R. T. (1975). Immunochemical isolation and characterization of albumin messenger ribonucleic acid. *J. Biol. Chem. 250*: 1759-1764.

Smith, K., Rosteck, Jr., P., and Lingrel, J. B. (1978). The location of the globin mRNA sequence within its 16S precursor. *Nucleic Acids Res. 5*: 105-115.

Taylor, J. M., and Hamilton, R. W. (1979). Affinity techniques for the isolation of specific mRNA and DNA sequences. In *Affinity Chromatography and Molecular Interactions*, J.-M. Egly (ed.). Editions INSERM, Paris, pp. 265-278.

Taylor, J. M., and Tse, T. P. H. (1976). Isolation of rat liver albumin messenger RNA. *J. Biol. Chem. 521*: 7461-7464.

Venetianer, P., and Leder, P. (1974). Enzymatic synthesis of solid phase-bound DNA sequences corresponding to specific mammalian genes. *Proc. Nat. Acad. Sci. U.S. 71*: 3892-3895.

Wood, T. G., and Lingrel, J. B. (1977). Purification of biologically active globin mRNA using cDNA-cellulose affinity chromatography. *J. Biol. Chem. 252*: 457-463.

Yogo, Y., and Wimmer, E. (1975). Sequence studies of poliovirus RNA. III. Polyuridylic acid and polyadenylic acid as components of the purified poliovirus replicative intermediate. *J. Mol. Biol. 92*: 467-477.

10
Lectins, Glycoproteins, Viruses, and Cells

GENERAL CONSIDERATIONS

Since the 1880s it has been known that extracts from certain plants have the property to agglutinate red blood cells. In the 1940s, agglutinins were discovered that are able to "select" types of cells based on their blood group activities. Although the term "lectin" (from Latin *legere*: to choose) was originally used to define agglutinins that could discriminate among types of red blood cells, today it is used more generally to include sugar binding proteins of different sources of origin, regardless of their ability to agglutinate cells. Such substances have been isolated from a wide variety of bacteria, yeasts, viruses, plants, invertebrates, and vertebrates. Their function in nature, however, is still unclear.

Most known lectins are multimeric, consisting of noncovalently associated identical (e.g., concanavalin A) or different subunits (e.g., *Ulex europeus* agglutinin). This multimeric structure is the cause of their ability to agglutinate cells or form precipitates with glycoconjugates in a manner similar to antigen-antibody interactions. Literally thousands of articles as well as various reviews on lectins and their application in affinity chromatography have been published (Boyd, 1962; Sharon and Lis, 1972; Lis and Sharon, 1973, 1977; Lotan and Nicolson, 1979; Sharon, 1979; Monsigny, 1979; Kristiansen et al., 1979). The statement of this matter, therefore, can be restricted here only to some selected examples.

A peculiarity of lectins is their ability to interact specifically and reversibly with carbohydrate groups or sequences. In this regard

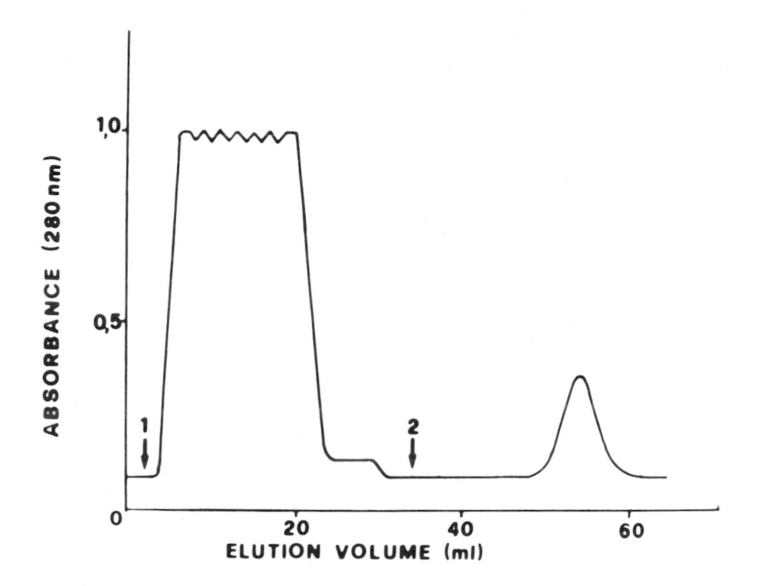

FIGURE 10.1 Affinity chromatography of *Medicago sativa* lectin on N-caproyl-D-galactosamine Sepharose. Arrow 1: Application of the sample in phosphate-buffered saline, pH 7.2. Arrow 2: Start of elution with 200 mM galactose in phosphate-buffered saline. (From Kamberger, 1978.)

they differ more or less from each other, as is shown for some examples in Table 5.2 of Chapter 5. The binding specificity of a given lectin can be easily determined in most cases by the inhibition of cell agglutination with simple sugars or oligosaccharides.

Since lectin-sugar interactions are reversible, lectins serve as ligands for affinity chromatographic separation of sugar groups or sequences containing biologically active substances. These are especially glycoproteins (inclusively surface glycoproteins from viruses and cell membranes), glycolipids and polysaccharides, enzyme-antibody and glycoprotein conjugates, as well as viruses, subcellular particles, and cells. Conversely, the affinity chromatographic separation of lectins is possible by adsorbent materials with suitable carbohydrate groups as accessible ligands.

SEPARATION OF LECTINS

Lectins can be separated from biological materials by affinity chromatography based on their specific binding properties to carbohydrates.

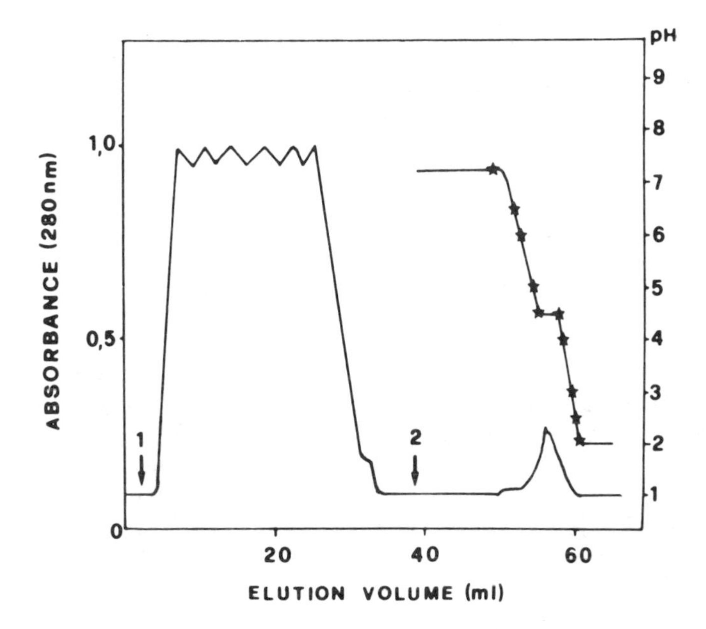

FIGURE 10.2 Affinity chromatography of *Medicago sativa* lectin on acid-treated Sepharose 6B. Arrow 1: Application of the sample in phosphate-buffered saline, pH 7.2. Arrow 2: Start of the elution with glycin·HCl buffer, pH 2.0. The line marked with a star shows the pH of the fractions. (From Kamberger, 1978.)

Adsorbents have been used mainly as (a) inhibiting carbohydrates immobilized to inert polymeric matrices (Lotan et al., 1973; Allen and Neuberger, 1974; Vretblad, 1976; Kieda et al., 1977); (b) pure, partially hydrolized (Kamberger, 1978), or cross-linked polysaccharides (Bywater et al., 1976; Bywater, 1978) with accessible carbohydrate residues that can act as ligands; (c) cross-linked cell fragments or cells containing the required carbohydrate groups as components of a membrane-fixed glycoprotein (Ochoa and Kristiansen, 1978; Lawny et al., 1978); and (d) synthetically prepared glycosides (Rafestin et al., 1974).

The elution of adsorbed lectins is carried out in most cases by buffered solutions of the specific carbohydrate derivative or by low-pH buffer. For example, the purification of the relatively unstable *Medicago sativa* lectin is shown in Figures 10.1 and 10.2 (Kamberger, 1978). In the first case N-caproyl-D-galactosamine-Sepharose 4B has been used as adsorbent material and the elution was done by a solution of 200 mM galactose in phosphate-buffered saline, pH 7.2. In the second case acid-treated Sepharose 6B beads served as adsorbent and glycin·HCl buffer, pH 2, as elution medium. Some further examples of affinity chromatographic lectin purification are given in Table 10.1.

AFFINITY CHROMATOGRAPHY OF GLYCOPROTEINS AND GLYCOPROTEIN CONJUGATES

Glycoproteins are proteins with covalently bound carbohydrate chains consisting of 1 or as many as 60 residues arranged in linear or branched structures (Finne et al., 1978). Immobilized lectins have been used successfully in the last few years to separate and purify numerous glycoproteins and glycoprotein conjugates. For example, immobilized concanavalin A has been proposed as a generally applicable adsorbent for purifying immunoglobulins (Weinstein et al., 1972), as well as brain and lysosomal hydrolases (Brattain et al., 1977). Further examples for glycoproteins separated by concanavalin A affinity chromatography have been described: α-fetoprotein (Pagé, 1973), horseradish peroxidase (Wagner, 1975), human alkaline (Trépanier et al., 1976) and acid phosphatase (van Etten and Saini, 1977), adenosine deaminase isoenzymes (Swallow et al., 1977), interferons of different origin (Davey et al., 1976; Berthold et al., 1978; Mizrahi et al., 1978), and rat uterine peroxidase (Keeping et al., 1981). Many other immobilized lectins have been also applied successfully for glycoprotein separation, and the number of publications on this matter increases from year to year.

Another field of application of lectin affinity chromatography is the purification of glycoprotein conjugates. As such, enzyme-labeled antibodies play an important role in enzyme immunoassays (see Chap. 8). Often for this reason peroxidase-immunoglobulin conjugates are used. These substances must be highly purified. Since both horseradish peroxidase and immunoglobulins are glycoproteins, immobilized lectins can be applied as adsorbent materials advantageously for this reason (Lannér et al., 1978; Varró and Barna-Vetro, 1978; Arends, 1981). As lectin, concanavalin A immobilized to Sepharose has been used and the elution carried out by buffered α-methyl-D-mannoside or methyl-α-D-glucopyranoside solution (Fig. 10.3).

MEMBRANE-BOUND GLYCOCONJUGATES

Lectin-mediated affinity chromatography has essentially contributed to the characterization of carbohydrate components of various viruses (Hayman et al., 1973; Kristiansen et al., 1979) and cells (Lotan and Nicholson, 1979). Cell membrane glycoconjugates – glycoproteins

TABLE 10.1 Lectins Purified by Affinity Chromatography

Lectin	Matrix	Elution conditions	References
Bandeiraea simplicifolia	Beaded cross-linked guaran	100 mM galactose for *E. lobata* seeds	Bywater et al. (1976)
Ricinus communis			
Echynocystis lobata seeds			
Wheat germ	Chitin	50 mM HCl	Block and Burger (1974)
Solanum tuberosum		100 mM acetic acid	Delmotte et al. (1975)
Peanut	Cross-linked erythrocyte membranes	100 mM D-galactose	Lawny et al. (1978)
Soybean		100 mM D-galactose	
Ulex europeus		50 mM fucose	
Sophora japonica		100 mM D-galactose	
Wheat germ		150 mM $(GlucNAc)_2$	

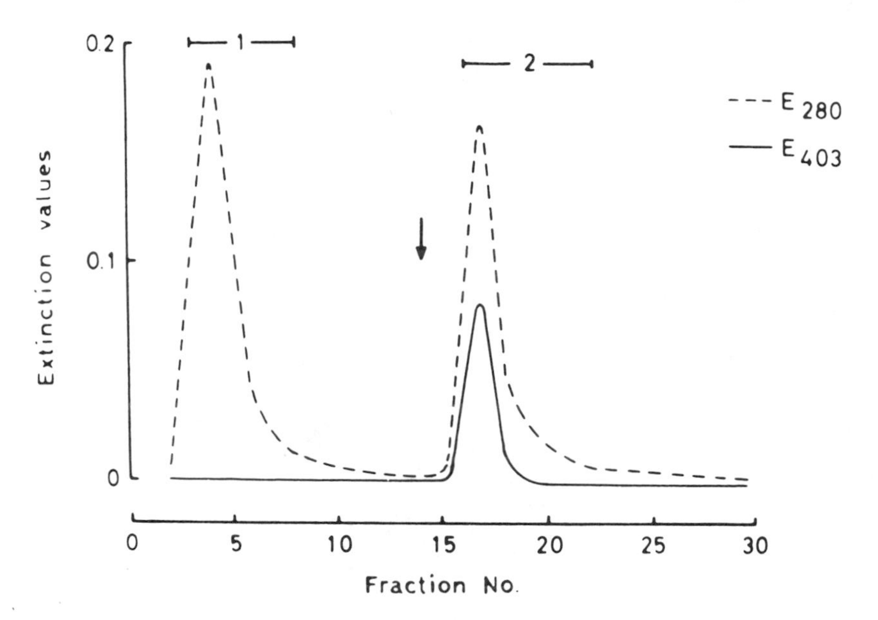

FIGURE 10.3 Affinity chromatography of peroxidase-labeled anti-human immunoglobulin on concanavalin A-Sepharose. The sample that contained 0.96 mg conjugate was applied to a 3 X 0.8 cm concanavalin A-Sepharose column. After washing, the adsorbed material was subsequently eluted by 10 mM α-methyl-D-mannoside, as indicated by the arrow. Only the peak 2 fractions were enzyme–immunologically active. (From Lánner et al., 1978.)

(Lotan and Nicolson, 1979) and glycolipids (Hakamori, 1975) – are inserted or intercalated to varying degrees into the bilayer, interrupting its continuity and forming mosaiclike arrangements (Nicolson, 1976). Membrane-fixed glycoconjugates are thought to play a vital role in many fundamental cellular processes, such as cell-cell recognition, transmission of extracellular stimuli, and regulation of cell movement, growth, division, and differentiation. The saccharides usually found in animal cell glycoproteins include N-acetylneuraminic acid (NeuNAc), galactose (Gal), mannose (Man), fucose (Fuc), N-acetylglucosamine (GlcNAc), and N-acetylgalactosamine (GalNAc). Figure 10.4 shows the general structure of the human erythrocyte membrane major sialoglycoprotein (glycophorin) (Kornfeld and Kornfeld, 1976).

In Chapter 11 the general principle of affinity chromatographic-glycoconjugate isolation is dealt with using the example of cell membrane-fixed glycoproteins. Their separation on immobilized lectins

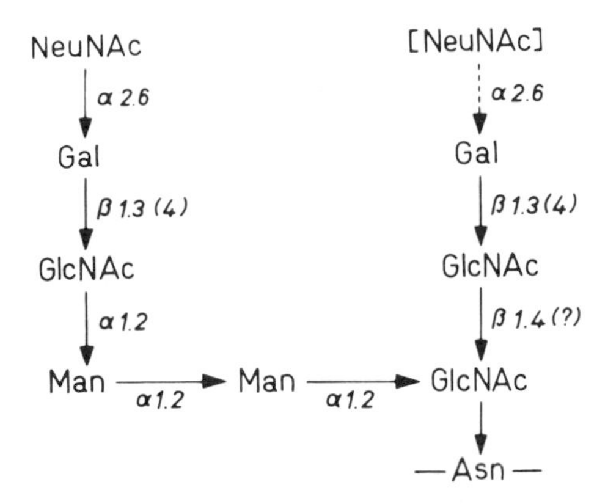

FIGURE 10.4 General structure of human erythrocyte membrane major sialoglycoprotein (glycophorin).

requires previous solubilization of the membrane. By far the most successful class of such agents are detergents (Helenius and Simmons, 1975). Since the separation procedures of solubilized integral membrane glycoproteins are usually carried out in the presence of the detergent, its effect on the adsorption of glycoproteins to the immobilized lectins must be considered. Extensive studies were conducted by Lotan and colleagues (1977), who have found that the nonionic detergents Nonidet P-40 and Triton X-100 do not remarkably affect the activities of the lectins up to concentrations approaching 2.5%, but considerable reduction in activity was observed with sodium dodecylsulfate. It is advisable in every case to test the influence of an applied detergent on affinity chromatographic efficiency before undertaking massive purification of membrane glycoconjugates. Despite these difficulties a great number of membrane glycoproteins have been isolated and characterized in the last few years (Lotan and Nicolson, 1979). Some selected examples are given in Table 10.2.

Since the affinity chromatographic isolation of native integral membrane proteins requires especially careful preparatory work, this matter of increasing interest in biochemistry is discussed more in detail in the following chapter.

AFFINITY CHROMATOGRAPHY OF VIRUSES

Since all enveloped viruses contain glycoproteins (Schwarz et al., 1977; Collins and Knight, 1978; Nakamura and Compans, 1978, 1979a, b), immobilized lectins have been applied in the isolation of

TABLE 10.2 Membrane Glycoproteins Isolated by Lectin Affinity Chromatography

Glycoprotein	Origin	Detergent	Immobilized lectin[a]	Eluting sugar	References
Glycophorin	Human red blood cells	Triton X-100	WGA	100 mM N-acetyl-glucosamine	Adair and Kornfeld (1974)
Band 3 glycoprotein					Findlay (1974)
Component 5.1	Rabbit thymocytes		ConA	1.6 mM α-methyl-D-glucopyranoside	Ullrich et al. (1978)
Thy-1-antigen	Rat thymocytes	Deoxycholate	LCA	500 mM methyl-α-D-glucopyranoside	Letarte-Muirhead et al. (1975)
	Rat brain			1 M methyl-α-D-glucopyranoside	Barclay et al. (1975)

Synaptic membrane glycoproteins	Rat brain cortex			2.5% N-acetylglycosamine	Gurd and Mahler (1974)
Membrane glycoproteins	Mouse Ehrlich ascites carcinoma cells		ConA (RCA, WGA)	100 mM α-methylmannoside (when ConA is used)	Nachbar et al. (1976)
Rhodopsin	Bovine retina	Dodecyltrimethylammonium bromide	ConA	100 mM glucose	Steinemann and Stryer (1973)
Membrane glycoprotein	Human blood platelets	Deoxycholate	LCA	2% (w/v) methyl-α-mannopyranoside	Clemetson et al. (1977)

[a] ConA: Concanavalin A; WGA: Wheat germ agglutinin; LCA: *Lens culinaris* agglutinin; RCA: *Ricinus communis* agglutinin.

these glycoproteins as well as in purification of intact viruses. Hayman and colleagues in 1973 succeeded in isolating viral glycoconjugates from influenza, fowl plague, and sendai viruses. More extensive investigations to purify viral glycoproteins and native viruses were performed by Kristiansen (1975) and Kristiansen and coworkers (1979). They found that the lectin of *Vicia ervilia* seeds (Fornstedt and Porath, 1975), which is specific for mannose and N-acetylglucosamine, can be applied for separation of intact influenza viruses but not for the glycoproteins after virus solubilization (e.g., by 0.2% w/v of the weakly anionic detergent sodium deoxycholate). In these experiments Sepharose 2B for lectin immobilization and a 2 M D-glucose solution for the desorption of bound viruses have been applied. On the other hand, *Crotolaria juncea* lectin columns were used successfully for the purification of bovine viral diarrhea virus and for the separation of influenza A virus glycoconjugates after solubilization (Kristiansen et al., 1979). Various experimentally obtained data, however, give rise to the assumption that hydrophobic interactions are involved in the virus lectin contacts.

AFFINITY CHROMATOGRAPHY OF CELLS

Separation of viable cells may be achieved by various methods, all resting on the biospecific recognition principle. It should be mentioned before embarking on a more detailed discussion that in every case a preliminary division by physical parameters, such as cell size or density, is advisable. For cell chromatography it is furthermore essential that (a) the applied adsorbent material is filled evenly into the column, leaving sufficient space between the beads to be able to pass through freely without being physically trapped, and (b) that these beads do not adsorb cells nonspecifically. These requirements are met, by agarose macrobeads with particle diameters of about 300 μm, among others.

1. The sugar specificity of lectins and their ability to interact reversibly with cell surface glycoconjugates has been used to fractionate cells (a) by differential agglutination with soluble lectins (Reisner et al., 1976, 1979) or (b) by specific adsorption to and elution from immobilized lectins (Sharon, 1979). Using such methods cell subpopulations may be distinguished that differ in their surface carbohydrate characteristics, as shown in Table 10.3 for some cell types fractionated by affinity chromatography.

Smith and colleagues (1978) have shown that differences in hydrophobic surface characteristics of porcine enteropathogenic *Escherichia coli* strains with or without K88 antigen can serve to differentiate between subpopulations by hydrophobic interaction chromatography on phenyl and octyl Sepharose CL 4B. In their experiments with paired

TABLE 10.3 Mammalian Cells Separated by Affinity Chromatography

Cells	Lectin[a]	Bound cells	Eluting sugar	References
Murine lymphocytes	PNA	Immature	D-galactose	Irlé et al. (1978)
Mouse spleen cells, treated with neuraminidase	HPA	T cells	N-acetyl-D-galactosamine	Haller et al. (1978)
Bone marrow cells	WGA	Stem enriched	N-acetyl-D-glucosamine	Nicola et al. (1978)
Peripheral blood lymphocytes treated with neuraminidase	HPA	T cells	N-acetyl-D-galactosamine	Hellström et al. (1976b)
T cells	WGA	Fractionation of two subpopulations	N-acetyl-D-glucosamine	Hellström et al. (1976a)

[a]PNA: Peanut agglutinin; HPA: *Helix pomatia* (snail) A hemagglutinin; WGA: wheat germ agglutinin.

K88-positive and K88-negative *E. coli* strains in buffered 4 M NaCl or 1 M $(NH_4)_2SO_4$ at 20°C (10 mM sodium phosphate buffer, pH 6.8), the K88-positive bacteria are strongly adsorbed whereas the K88-negative subpopulation did not appear to adsorb either gel. Release of bacteria that adsorbed to the hydrophobic gels was possible by decreasing the ionic strength. The authors hold the opinion that such hydrophobic interactions also play a role in cell separation by immobilized lectins.

2. Repeatedly, antigen-antibody interactions have been utilized for cell separation. A method for the preparation of highly purified human T- and B-lymphocyte subpopulations using anti-F_{ab} antibodies covalently coupled to Sephadex G200 was described by Chess and coworkers (1974). Columns of anti-rabbit $F_{ab'}$-antibodies immobilized to the macrobeaded Sepharose 6MB were applied successfully to prepare T and B memory cells from lymph nodes of rabbits preimmunized with keyhole limpet hemocyanin (Manderino et al., 1978).

Often, however, antigen-antibody interactions are so strong (association constant $>10^9 M^{-1}$) that an elution of the bound cells is possible only under conditions leading to a loss of their viability (such as extreme pH, denaturing agents, or chaotropic salts; see Chap. 8). In such cases protein A as ligand coupled to macrobeads (e.g., Sepharose 6MB) represents an alternative method for purifying diverse cell types. Protein A itself is a protein isolated from *Staphylococcus aureus* that binds specifically to the Fc region of immunoglobulins (see Fig. 8.1, Chap. 8) (Kronvall et al., 1970). When this protein is covalently coupled to macrobeads it can be used advantageously to adsorb cells that have been coated with a specific antibody. Using an excess of soluble IgG class antibodies, the antibody-coated cells may be displaced again (Ghetie et al., 1978).

3. Although cell separation by immobilized lectins and antibodies (or protein A) has been most frequently applied, in some cases a specific and reversible cell surface receptor-ligand interaction can be utilized successfully for this reason. For example, cell separation by such a procedure has been described using the purification of acetylcholine receptor-bearing neuronal cells from preparations of chick embryo sympathetic ganglion cells by snake venom α-bungarotoxin immobilized on Sepharose 6MB (Dvorak et al., 1978). It has been found that the adsorbed neuronal cells were not released by competitive elution but instead by trypsin digestion. Although this treatment presumably attacks protein molecules on the cell surface, the cells obtained in this way were viable.

REFERENCES

Adair, W. L., and Kornfeld, S. (1974). Location of the receptors of wheat germ agglutinin and the ricinus communis lectins from human erythrocytes using affinity chromatography. *J. Biol. Chem. 249*: 4696-4704.

Allen, A. K., and Neuberger, A. (1974). A simple method for the preparation of an adsorbent for soybean agglutinin using galactosamine and CH-Sepharose. *FEBS Lett. 50*: 362-364.

Arends, J. (1981). Purification of peroxidase-conjugate antibody for enzyme immunoassay by affinity chromatography on concanavalin A. *Methods Enzymol. 63*: 116-175.

Barcley, A. N., Letarte-Muirhead, M., and Williams, A. F. (1975). Purification of the Thy-1 molecule from rat brain. *Biochem. J. 151*: 699-706.

Berthold, W., Tan, C., and Tan, Y. H. (1978). Purification and in vitro labeling of interferon from human fibroblastoid cell line. *J. Biol. Chem. 253*: 5206-5212.

Block, R., and Burger, M. M. (1974). Purification of wheat germ agglutinin using affinity chromatography on chitin. *Biochem. Biophys. Res. Commun. 58*: 13-19.

Boyd, W. C. (1962). *Introduction to Immunochemical Specificity*. Wiley-Interscience, New York.

Brattain, M. G., Kimball, P. M., Pretlow, II, T. G., and Marks, M. E. (1977). The interaction of N-acetylhexosamidase with insolubilized concanavalin A. *Biochem. J. 163*: 247-251.

Bywater, R. (1978). Purification of lectins on beaded polysaccharide materials. In *Chromatography of Synthetic and Biological Polymers*, *Vol. 2*, R. Epton (ed.). Ellis Horwood, Chichester, pp. 325-329.

Bywater, R., Goldstein, I. J., and Lönngren, J. (1976). Cross-linked guaran: A versatile immunosorbent for D-galactopyranosyl binding lectins. *FEBS Lett. 68*: 31-34.

Chess, L., McDermott, R. P., and Schlossmann, S. F. (1974). Immunologic functions of isolated human lymphocyte subpopulations. I. Quantitative isolation of human T and B cells and response to mitogens. *J. Immunol. 113*: 1113-1121.

Clemetson, K. J., Pfueller, S. L., Luscher, E. F., and Jenkins, C. S. P. (1977). Isolation of the membrane glycoproteins of human blood platelets by lectin affinity chromatography. *Biochim. Biophys. Acta 464*: 493-508.

Collins, I. K., and Knight, C. A. (1978). Purification of the influenza virus glycoprotein and characterization of its carbohydrate components. *J. Virol. 26*: 457-467.

Davey, M. W., Sulkowski, E., and Carter, W. A. (1976). Binding of human fibroblast interferon to concanavalin A agarose. Involvement of carbohydrate recognition and hydrophobic interaction. *Biochemistry 15*: 704-713.

Delmotte, F., Kiéda, C., and Monsigny, M. (1975). Protein sugar interaction: Purification by affinity chromatography of Solanum tuberosum agglutinin (STA lectin). *FEBS Lett. 53*: 324-330.

Dvorak, D. J., Gipps, E., and Kidson, C. (1978). Isolation of specific neurons by affinity methods. *Nature 271*: 564-566.
Findlay, J. B. C. (1974). The receptor protein for concanavalin A and Lens culinaris phytohemagglutinin in the membrane of human erythrocyte. *J. Biol. Chem. 249*: 4398-4403.
Finne, J., Krusius, T., Rauvala, H., Kekomaki, R., and Myllyla, G. (1978). Alkali stable blood group A- and B-active poly(glycosyl)-peptides from human erythrocyte membrane. *FEBS Lett. 89*: 111-115.
Fornstedt, N., and Porath, J. (1975). Characterization studies on a new lectin found in seeds of Vicia ervilia. *FEBS Lett. 57*: 187-191.
Ghetie, V., Mota, G., and Sjöquist, J. (1978). Separation of cells by affinity chromatography on SpA-Sepharose 6 MB. *J. Immunol. Methods 21*: 133-141.
Gurd, J. W., and Mahler, H. R. (1974). Fractionation of synaptic plasma membrane glycoproteins by lectin affinity chromatography. *Biochemistry 13*: 5193-5198.
Hakamori, S.-J. (1975). Structures and organization of cell surface glycolipids dependency on cell growth and malignant transformation. *Biochim. Biophys. Acta 417*: 55-89.
Haller, O., Gidlung, M., Hellström, U., Hammarström, S., and Wigzell, H. (1978). A new surface marker on mouse natural killer cells: Receptors for Helix pomatia A agglutinin. *Eur. J. Immunol. 8*: 765-771.
Hayman, M. J., Skehel, J. J., and Crumpton, M. J. (1973). Purification of virus glycoproteins by affinity chromatography using Lens culinaris phytohaemagglutinin. *FEBS Lett. 29*: 185-188.
Helenius, A., and Simmons, K. (1975). Solubilization of membranes by detergents. *Biochim. Biophys. Acta 415*: 29-79.
Hellström, U., Dillner, M.-L., Hammarström, S., and Perlmann, P. (1976a). The interaction of non-mitogenic and mitogenic lectins with T lymphocytes: Association of cellular receptor sites. *Scand. J. Immunol. 5*: 45-54.
Hellström, U., Hammarström, S., Dillner, M.-L., Perlmann, H., and Perlmann, P. (1976b). Fractionation of human blood lymphocytes on Helix pomatia A hemagglutinin coupled to Sepharose beads. *Scand. J. Immunol. 5, Suppl. 5*: 45-55.
Irlé, J., Piguet, P.-F., and Vassali, P. (1978). In vitro maturation of immature thymocytes into immunocompetent T cells in absence of direct thymic influence. *J. Exp. Med. 148*: 32-45.
Kamberger, W. (1978). Binding specificity and purification of Medicago sativa lectin. In *Affinity Chromatography*, O. Hoffmann-Ostenhof, M. Breitenbach, F. Koller, D. Kraft, and O. Scheiner (eds.). Pergamon Press, Oxford, pp. 295-298.
Keeping, H. S., Kimura, S., Lovsted, J., and Jellinck, P. H. (1981). Further purification and properties of rat uterine peroxidase. *Can. J. Biochem. 59*: 916-920.

Kieda, C., Delmotte, F., and Monsigny, M. (1977). Protein-sugar interactions: Preparation, purification and properties of rabbit antibodies against di-N-acetylchitobiose. *Proc. Nat. Acad. Sci. U.S. 74*: 168-172.

Kornfeld, R., and Kornfeld, S. (1976). Comparative aspects of glycoprotein structure. *Annu. Rev. Biochem. 45*: 217-237.

Kristiansen, T. (1975). Virus purification by Vicia ervilia lectin coupled to Sepharose. *Prot. Biol. Fluids 23*: 663-665.

Kristiansen, T., Sparrman, M., and Moreno-Lopez, J. (1979). Immobilized lectins for isolation of virus and viral glycoproteins. In *Affinity Chromatography and Molecular Interactions*, J. Egly (ed.). Editions INSERM, Paris, pp. 217-229.

Kronvall, G., Seal, U. S., Finstead, S., and Williams, Jr., R. C. (1970). Phylogenetic insight into the evolution of mammalian fragment γG globulin using staphylococcal protein A. *J. Immunol. 104*: 140-147.

Lannér, M., Bergquist, R., Carlsson, J., and Huldt, G. (1978). Purification of enzyme labeled conjugate by affinity chromatography. In *Affinity Chromatography*, O. Hoffmann-Ostenhof, M. Breitenbach, F. Koller, D. Kraft, and O. Scheiner (eds.). Pergamon Press, Oxford, pp. 237-241.

Lawny, L., Bot, M. H., Lentwojt, E., and Segard, E. (1978). Cross-linked erythrocyte membrane columns as a tool for affinity chromatography of lectin. In *Affinity Chromatography*, O. Hoffmann-Ostenhof, M. Breitenbach, F. Koller, D. Kraft, and O. Scheiner (eds.). Pergamon Press, Oxford, pp. 299-302.

Letarte-Muirhead, M., Barclay, A. N., and Williams, A. F. (1975). Purification of the thy-1-molecule, a major cell-surface glycoprotein of rat thymocytes. *Biochem. J. 151*: 685-697.

Lis, H., and Sharon, N. (1973). The biochemistry of plant lectins (phytohemagglutinins). *Annu. Rev. Biochem. 42*: 541-574.

Lis, H., and Sharon, N. (1977). Lectins: Their chemistry and application in immunology. In *The Antigens*, Vol. 4, M. Sela (ed.). Academic Press, New York, pp. 429-529.

Lotan, R., and Nicolson, G. L. (1979). Purification of cell membrane glycoproteins by lectin affinity chromatography. *Biochim. Biophys. Acta 559*: 329-376.

Lotan, R., Gussin, A. E. S., Lis, N., and Sharon, N. (1973). Purification of Wheat germ agglutinin by affinity chromatography on Sepharose bound N-acetylglucosamine derivative. *Biochem. Biophys. Res. Commun. 52*: 656-662.

Lotan, R., Beattie, G., Hubbell, W., and Nicolson, G. L. (1977). Activities of lectins and their immobilized derivatives in detergent solutions. Implication on the use of lectin affinity chromatography for the purification of membrane glycoproteins. *Biochemistry 16*: 1787-1789.

Manderino, G. L., Gooch, G. T., and Stavitsky, A. B. (1978). Preparation, characterization and functions of rabbit lymph node cell populations. I. Preparation of KLH primed T and B memory cells with anti-$F_{ab'}$ affinity columns. *Cell. Immunol. 41*: 264-275.

Mizrahi, A., O'Malley, J. A., Carter, W. A., Takatsuki, A., Tamura, G., and Sulkowski, E. (1978). Glycosylation of interferons. Effects of tunicamycin on human immune interferon. *J. Biol. Chem. 253*: 7612-7615.

Monsigny, M. (1979). Protein sugar interactions: Membrane glycoconjugates and membrane lectins. In *Affinity Chromatography and Related Techniques*, J.-M. Egly (ed.). Editions INSERM, Paris, pp. 207-215.

Nachbar, M. S., Oppenheim, J. D., and Aull, F. (1976). Interactions of lectins with plasma membrane glycoproteins of the Ehrlich ascites carcinoma cells. *Biochim. Biophys. Acta 419*: 512-529.

Nakamura, K., and Compans, R. W. (1978). Glycopeptide components of influenza viral glycoproteins. *Virology 86*: 432-442.

Nakamura, K., and Compans, R. W. (1979a). Biosynthesis of the oligosaccharides of influenza viral glycoproteins. *Virology 93*: 31-47.

Nakamura, K., and Compans, R. W. (1979b). Host-cell and virus strain dependent differences in oligosaccharides of hemagglutinin glycoproteins of influenza A viruses. *Virology 95*: 8-23.

Nicola, A. N., Burgess, A. W., Metcalf, D., and Battye, F. L. (1978). Separation of mouse bone marrow cells using wheat germ agglutinin affinity chromatography. *Aust. J. Exp. Biol. Med. Sci. 56*: 663-679.

Nicolson, G. L. (1976). Transmembrane control of the receptors on normal and tumor cells. I. Cytoplasmic influence over all surface components. *Biochim. Biophys. Acta 457*: 57-108.

Ochoa, J.-L., and Kristiansen, T. (1978). Stroma: As an affinity adsorbent for non-inhibitable lectins. *FEBS Lett. 90*: 145-148.

Pagé, M. (1973). α-Foetoprotein: Purification on Sepharose linked concanavalin A. *Can. J. Biochem. 51*: 1213-1215.

Rafestin, M. E., Obrenovitch, A., Oblin, A., and Monsigny, M. (1974). Purification of N-acetyl-D-glucosamine binding proteins by affinity chromatography. *FEBS Lett. 40*: 62-66.

Reisner, Y., Ravid, A., and Sharon, N. (1976). Use of soybean agglutinin for the separation of mouse B and T lymphocytes. *Biochem. Biophys. Res. Commun. 72*: 1585-1591.

Reisner, Y., Biniaminov, M., Rosenthal, E., Sharon, N., and Ramot, B. (1979). Interaction of peanut agglutinin with normal human lymphocytes and with leukemic cells. *Proc. Natl. Acad. Sci. U.S. 76*: 447-451.

Schmidt-Ullrich, R., Mikkelsen, R. B., and Wallach, D. F. H. (1978). Transmembrane disposition of the 55000 Dalton concanavalin A receptor protein of thymocyte plasma membranes. *J. Biol. Chem.* *253*: 6973-6978.

Schwarz, R. T., Schmidt, M. F. G., Anwer, U., and Klenk, H.-D. (1977). Carbohydrates of influenza viruses. I. Glycopeptides derived from viral glycoproteins after labeling with radioactive sugars. *J. Virol.* *23*: 23-30.

Sharon, N. (1979). Application of lectins to cell fractionation. In *Affinity Chromatography and Molecular Interactions*, J.-M. Egly (ed.). Editions INSERM, Paris, pp. 197-205.

Sharon, N., and Lis, H. (1972). Lectins: Cell agglutinating and sugar specific proteins. *Science* *177*: 949-959.

Smith, C. J., Jonsson, P., Olsson, E., Söderlind, O., Rosengren, J., Hjerten, S., and Waldström, T. (1978). Differences in hydrophobic surface characteristics of porcine enteropathogenic Escherichia coli with or without K88 antigen as revealed by hydrophobic interaction chromatography. *Infect. Immun.* *22*: 462-472.

Steinemann, A., and Stryer, L. (1973). Accessibility of carbohydrate moiety of rhododopsin. *Biochemistry* *12*: 1499-1502.

Swallow, D. M., Evans, L., and Hopkinson, D. A. (1977). Several of the adenosine deaminase isoenzymes are glycoproteins. *Nature* *269*: 261-262.

Trepanier, J. M., Seargant, L. E., and Stinson, R. A. (1976). Affinity purification and some molecular properties of human liver alkaline phosphatase. *Biochem. J.* *155*: 653-660.

Van Etten, R. L., and Saini, M. S. (1977). Preparation of homogeneous human prostatic acid phosphatase using concanavalin A-Sepharose 4B. *Biochim. Biophys. Acta* *484*: 487-492.

Varró, R., and Barna-Vetro, I. (1978). The use of affinity chromatography for purification of enzyme antibody conjugates. In *Affinity Chromatography*, O. Hoffmann-Ostenhof, M. Breitenbach, F. Koller, D. Kraft, and O. Scheiner (eds.). Pergamon Press, Oxford, pp. 247-250.

Vretblad, P. (1976). Purification of lectins by biospecific affinity chromatography. *Biochim. Biophys. Acta* *434*: 169-176.

Wagner, M. (1975). Reinigung von Meerrettich Peroxidase durch Affinitätschromatographie an Concanavalin A-agarose. *Acta Biol. Med. Germ.* *34*: 1429-1431.

Weinstein, Y., Givol, D., and Strausbauch, P. H. (1972). The fractionation of immunoglobulins with insolubilized concanavalin A. *J. Immunol.* *109*: 1402-1404.

11

Integral Membrane Proteins
Amphiphatic Nature, Solubilization, and Peculiarities in their Purification by Affinity Methods

W.-H. Schunck
Central Institute of Molecular Biology, Academy of Sciences of the GDR

Membrane proteins are involved as receptors, carriers, electron transport components, and enzymes in many of the most fundamental processes of the living cell (for a functional classification, see Freedmann, 1981). However, they have been less characterized than soluble proteins due to particular difficulties in isolating and assaying them, retaining their native structure, and reconstituting their original functional properties after purification.

Comparing the investigations of membrane proteins as summarized by Guidotti (1972) and Coleman (1973) with that represented by some more recent reviews (Capaldi, 1977; Finean and Michell, 1981; Brock and Tanner, 1982), it is obvious that considerable progress has been made in the last decade, mostly with respect to

- The discovery of several fundamental principles for structure and organization of biological membranes (Singer and Nicolson, 1972; Singer, 1974, 1977; Israelachvilli et al., 1980; Tanford, 1980; Finean and Michel, 1981)
- The elucidation of structural peculiarities of integral membrane proteins at the primary sequence level (von Heijne, 1981; Warren, 1981, Capaldi, 1982)
- The concept of protein topogenesis, assuming a "code" in the form of discrete sequences of the polypeptide chain (Blobel, 1980;

Kreil, 1981; Brock and Tanner, 1982), which determines the final location of proteins

The introduction of nondenaturating detergents as a general approach for solubilizing membrane proteins with retention of activity (Helenius and Simons, 1975; Tanford and Reynolds, 1976)

The development of techniques for reconstituting purified membrane proteins into liposomes (Szoka and Papahadjopoulos, 1980) to imitate the microenvironment originally provided by the biomembrane (Gennis and Jonas, 1977; Racker, 1979; Chapman, 1982; Eyton, 1982)

The improvement of analytic techniques for studying the protein pattern of membranes (O'Farell, 1975; Azzi et al., 1981; Finean and Michell, 1981)

On the basis of this knowledge rapid further progress in isolation and characteriazation of a great number of membrane proteins can be expected in the next few years. Thereby, affinity chromatography offers in many cases the only promising approach for purification of such biosubstances. However, its successful application in this field is largely dependent on adequate handling of this peculiar group of proteins. In this respect, some useful generalizations can be derived from our present knowledge about the structure and function of proteins in biological membranes, which is the main aim of this chapter.

CLASSIFICATION OF MEMBRANE PROTEINS

The proteins found in association with membranes have been classified into two categories, integral (or intrinsic) and peripheral (or extrinsic) (Singer, 1974). The integral membrane proteins traverse the hydrocarbon core of the phospholipid bilayer or are firmly anchored in it, mainly via hydrophobic interactions (Tanford, 1980). Consequently, they can be dissociated in general only by disrupting the membrane structure. They are poorly or not water soluble, form aggregates with detergents or lipids, and often tend to precipitate. Therefore, special care must be taken to keep them in solution and to retain their native structure throughout a purification procedure.

In contrast, the so-called peripheral membrane proteins, such as cytochrome c and spectrin, are bound to the polar membrane surface, formed by the head groups of lipid molecules and the hydrophilic domains of integral membrane proteins. They are dislodged even by relatively mild treatments and may be handled like ordinary water-soluble proteins in further purification steps. Their content in membrane preparations depends on the conditions used (ionic strength, pH, presence of divalent cations and others), and they represent probably an artificial residue of much larger functional complexes, existing in vivo between integral membrane constituents and the adjacent cytoplasm (Nicolson, 1976; Geiger, 1983).

MOLECULAR PROPERTIES

Our present knowledge about the structure of integral membrane proteins has been derived mainly from extensive studies on bacteriorhodopsin (Engelman et al., 1980), rhodopsin (Dratz and Hargrave, 1983), glycophorin A (Tomita and Marchesi, 1975), cytochrome $\underline{b}_5$ (Spatz and Strittmatter, 1971; Bendzko and Pfeil, 1983), cytochrome oxidase (Fuller et al., 1979), and on a few others; for reviews, see Warren (1981), Capaldi (1982), and Brock and Tanner (1982).

It is generally assumed that the decisive difference between integral membrane proteins and water-soluble proteins is to be found at the primary sequence level and that the integration of a certain protein into the lipid bilayer is predominantly due to the thermodynamically favorable removal of long stretches of apolar amino acids from contact with water (Tanford, 1980). The presence of such membrane-penetrating segments, formed by about 19-23 uncharged amino acids, has been proved for several integral membrane proteins (von Heijne, 1981). These proteins traverse the 30 Å hydrocarbon core of the biomembrane probably by single α helices and have hydrophilic domains of different size that are projected outside the bilayer. Depending on the topology of the hydrophilic domains, being exposed only on one or on both sides of the membrane, integral membrane proteins can be classified as monotopic, bitopic, and polytopic (Blobel, 1980). Monotopic proteins, like cytochrome b_5, are anchored to the membrane only by a hydrophobic sequence, which probably does not completely traverse the hydrocarbon core of the bilayer. Bitopic proteins span the bilayer once and contain hydrophilic domains on both sides of the membrane (e.g., glycophorin A). Polytopic proteins contain several membrane-penetrating hydrophobic segments. To this group belong, besides the well-known bacteriorhodopsin, the cytochrome P_{450}LM2 (Heinemann and Ozols, 1983) and rhodopsin (Dratz and Hargrave, 1983), as derived from their recently elucidated amino acid sequences.

These examples illustrate one important structural feature, probably shared by all integral membrane proteins — they all must be assumed to consist of at least two different domains: a hydrophilic and a hydrophobic. This means integral membrane proteins belong to the amphiphilic compounds (for classification, see Helenius and Simons, 1975; Lichtenberg et al., 1983).

Their solubility depends on the balance of "size and strength" of the two contrary domains. In general, neither pure water nor a pure organic solvent will be an appropriate medium. There are exceptions representing the possible extremes: on the one hand, the monotopic proteins with large hydrophilic domains (for references, see Helenius and Simons, 1975), which are water-soluble by forming defined aggregates in which the hydrophobic parts are removed from contact with water, and, on the other hand, the highly hydrophobic proteolipids (Lees et al., 1979; Schlesinger, 1981), which will partition into the organic phase in a water-organic solvent two-phase system.

SOLUBILIZATION

Since the solubilization of membranes represents the most crucial step in any attempt to purify an integral membrane protein, it is discussed in detail in the following. Among the different techniques available, the use of detergents proves to be the most generally useful method (Helenius and Simons, 1975). Authorative reviews have been published concerning the use of detergents for solubilization of membrane constituents (Helenius and Simons, 1975; Lichtenberg et al., 1983), the characterization of membrane proteins in detergent solutions (Tanford and Reynolds, 1976), the properties of detergents (Helenius et al., 1979), the removal of unbound detergent from membrane proteins (Furth, 1980), and the physical fundamentals involved (Tanford, 1980).

According to their mode of interaction with proteins, they can be roughly divided into denaturating and "mild" detergents. For denaturating detergents, like sodium dodecylsulfate (Reynolds and Tanford, 1970) and other long-chain ionic alkyl derivatives (Nozaki et al., 1974), cooperative binding of large numbers of detergent molecules was observed with all soluble as well as membrane proteins studied so far. Mostly, they cause gross structural alterations in proteins accompanied with a loss of biological activity. The soluble complexes formed are, however, especially suitable for the application of some analytic techniques, such as polyacrylamide gel electrophoresis (Weber et al., 1969; Laemmli, 1970; O'Farell, 1975).

In contrast, "mild detergents," like the bile salts cholate, deoxycholate, and digitonine and nonionic detergents, for example, of the Triton and Tween series (for structures and properties, see Helenius et al., 1979), are not usually bound to ordinary water-soluble and peripheral membrane proteins. The binding of mild detergents was found to be a specific feature of integral membrane proteins (Helenius and Simons, 1972; Makino et al., 1973; Clarke, 1975) and additionally of some special serum proteins with high-affinity sites for lipid binding.

It is generally assumed that the bound mild detergents form micelle-like clusters around the hydrophobic domains of the integral membrane proteins, thereby retaining in most cases the native protein structure. In this respect, cytochrome $\underline{b}_5$ of the endoplasmic reticulum provides an example. It consists of a large globular hydrophilic domain exposed to the cytoplasma and a hydrophobic segment anchored to the membrane. The two domains can be split proteolytically, and in this way, using radioactive labeled Triton X-100 and deoxycholate, it was demonstrated that the hydrophobic part of the protein is exclusively responsible for detergent binding (Robinson and Tanford, 1975). Usually, 80-100 mol of detergent are bound per mole of protein, corresponding to 0.3-1.5 mg/mg on a weight per weight basis for the complexes studied between Triton X-100 and different membrane proteins (for references, see Helenius and Simons, 1975; Gennis and Jonas, 1977; Capaldi, 1977). Higher binding ratios in connection with denaturation of the hydrophilic domains are prevented in the case of mild detergents, presumably because of their relatively low critical micelle concentrations, which automatically limit the free monomer concentrations (Makino et al., 1973).

The extraction of intact proteins from biological membranes takes advantage of these properties of the mild detergents. The exact sequence of events in the solubilization process has been studied with artificial phospholipid vesicles (Lichtenberg et al., 1983) and with two relatively simple membrane systems: the Semliki Forest virus membrane containing three different spike proteins (Helenius and Simons, 1975; Simons et al., 1977) and the sarcoplasmic recticulum containing the Ca^{2+}-ATPase (Prado et al., 1983).

Accordingly, the dissociation of simple biological membranes by in creasing detergent concentrations can be described by the following steps:

1. At low detergent concentration (up to nearly a 1:1 molar detergent phospholipid ratio), detergent monomers are incorporated into the membrane, leading to an increased membrane permeability.
2. When the detergent-lipid ratio exceeds a critical value (1.5:1 in the case of the sarcoplasmic reticulum treated with Triton X-100), membrane vesicles are transformed into mixed lipid-protein-detergent micelles.
3. At even higher detergent concentrations, the large lipid-protein-detergent micelles are resolved under formation of protein-detergent micelles, lipid-detergent micelles, and pure detergent micelles.

This scheme includes that the lipid molecules surrounding the hydrophobic domains of integral membrane proteins are replaced by detergent molecules, so that their two phase orientation and native structure can be generally preserved in the solubilization process (Robinson and Tanford, 1975; Uterman and Simons, 1974).

With mitochondrial and other more complex biological membranes, step 2 proceeds in a modified manner: there is no critical detergent-lipid ratio common to the solubilization of all the membrane constituents (Barbero et al., 1982; MacDonald, 1980). This phenomenon allows a "selective solubilization" (Helenius and Simons, 1975) of certain integral membrane proteins, as has been applied for example in the purification of lactose permease from *Escherchia coli* (Newman et al., 1981) and (Na^+, K^+)-ATPase from plasma membranes (Kracke et al., 1981).

Accordingly, the following procedure should be tested, to achieve a remarkable enrichment of the integral membrane protein even in the solubilization step:

Removal of peripheral membrane proteins by several washing steps using buffers with different ionic strength, pH values, and chelating agents

Attempts at selective solubilization by using several mild detergents that differ in hydrophilic-lipophilic balance values (usually in the range 15-12), by applying different detergent concentrations or by changing some general conditions, such as ionic strength.

The choice of the appropriate detergent is still largely a matter of trial and error. Most workers have used nonionic detergents of the phenylalkylpolyoxyethylene type and bile salts (see Table 11.1). The main difference between these two groups appears to consist in their effects on protein-protein interactions. Whereas quarternary structures are preserved in many cases using nonionic detergents, bile salts are most suited to achieve mixed micelles in which monomeric proteins are enclosed (Simons et al., 1977; Rizzolo, 1981).

AFFINITY METHODS

Table 11.1 illustrates that affinity chromatography has turned out to be an extremely useful technique for the purification of numerous integral membrane proteins. In general, four decisive points must be considered in the elaboration of such purification procedures:

1. The solubilization with retention of native structure
2. The establishment of an assay method applicable throughout the purification steps (especially in the case of vectorially acting enzymes and carrier proteins)
3. The development of an appropriate affinity resin
4. The reconstitution of the in situ functional properties, usually by incorporation of the purified protein into artificial lipid vesicles (Eyton, 1982).

All the purification procedures listed take advantage of the mechanism of detergent solubilization already described, and the proteins are

TABLE 11.1 Selected Examples for the Application of Affinity Chromatography to the Purification of Integral Membrane Proteins

Protein	Source	Detergent used for solubilization	Immobilized ligand	References
1. Receptors				
Insulin	Liver membranes	Triton X-100	Insulin	Cuatrecasas and Parikh (1974)
	Placental membranes	Triton X-100	Wheat germ agglutinin, insulin	Fujita-Yamaguchi (1983)
Prolactin	Mammary gland	Triton X-100	Human growth hormone	Shiu and Friesen (1974)
Growth hormone	Liver membranes	Triton X-100	Human growth hormone	Waters and Friesen (1978)
Nicotinic acetylcholine	Electric organs, mammalian muscles	Triton X-100	Cholinergic ligands or α-neurotoxin	Heidman and Changeux (1978); Conti-Tronconi and Raftery (1982)
α_1-Adrenergic	Liver membranes	Digitonin	Analog of prazosin	Graham et al. (1982)
β-Adrenergic	Erythrocyte membranes	Digitonin	Alprenolol	Cerione et al. (1983)
γ-Aminobutyric acid	Brain postsynaptic membranes	Deoxycholate	Benzodiazepine	Sigel et al. (1983)

TABLE 11.1 (continued)

Protein	Source	Detergent used for solubilization	Immobilized ligand	References
Glycine	Spinal cord synaptic membranes	Triton X-100	Aminostrychnine, wheat germ agglutinin	Pfeiffer et al. (1982)
Serum glycoprotein				For a review, see Ashwell and Harford (1982)
Desialylated	Liver membranes	Triton X-100	Asialoporosomucoid	Pricer et al. (1974)
With terminal mannose residues	Liver membranes	Triton X-100	Mannan	Kawasaki et al. (1978)
Complement (C3b component)	Macrophage membranes	NP 40	C3b-component of complement	Schneider et al. (1981)
2. Carrier				
Glutamate	Heart mitochondria	Cholate	Glutamate	Gautheron and Julliard (1979)
Na^+/D-glucose cotransporter	Brush-border membranes	Digitonin	Monoclonal antibodies from hybridome screening	Schmidt et al. (1983)
Na^+-independent glucose transporter	Jejunal basalolateral membranes	Cholate	Castor bean lectin	Ling and Faust (1983)

3. Electron transfer proteins and enzymes				
Cytochrome c oxidase	Mitochondria	Triton X-100	Cytochrome c	Bill et al. (1980)
		Cholate-Triton X-100		Rascati and Parsons (1979)
		Laurylmaltoside		Thompson and Ferguson-Miller (1983)
Cytochrome P_{450}	Adrenal mitochondria	Cholate	Adrenodoxin	Akhrem et al. (1979)
	Liver microsomes	Cholate	1,8-Diamino-octan	Imai and Sato (1974)
	Microsomes of alkane-grown yeast	Cholate	1,8-Diamino-octan	Riege et al. (1981); Schunck et al. (1983)
NADPH-cytochrome P_{450} reductase	Liver microsomes	Renex 690-cholate	2',5'-ADP	Yasukochi and Masters (1976)
		Cholate	2',5'-ADP	Shepard et al. (1983)
Phospholipase A2	Erythrocyte membranes	Cholate	Lecitin-ether analog	Zahler et al. (1981)
Acetylcholine esterase	Erythrocyte membranes	Triton X-100	Quaternary nitrogen aromatic compound	Brodbeck et al. (1981)
Ca^{2+}-ATPase	Heart sarcolemma	Triton X-100	Calmodulin	Caroni and Carafoli (1981)
	Liver microsomes	Triton X-100	Calmodulin	Moore and Kraus-Friedman (1983)
α-Mannosidase	Liver microsomes	Triton X-100	Concanavalin A	Forsee and Schutzbach (1981)
γ-Glutamyltransferase	Hepatoma membranes	Triton X-100	Antibodies against the kidney isozyme	Tsuchida and Sato (1983)

kept in a soluble state throughout the chromatographic steps by the presence of sufficiently high detergent concentrations. Bile salts have been used advantageously to resolve multicomponent enzyme systems, for example the cytochrome P_{450} systems of different origin, which catalyze the monoxygenase transformation of several lipophilic substrates. They consist of cytochrome P_{450} as the terminal oxidase and at least one or two electron transfer proteins (for a review, see Sato and Omura, 1979). The components were purified separately after cholate solubilization (Table 11.1) and reconstituted to active enzyme systems in the presence of low detergent concentrations or lipids. In contrast, nonionic detergents are commonly used in those cases in which a loss of non-ligand-binding subunits or components should be avoided in the affinity chromatographic steps, as in the purification of hormone receptors.

Mostly, selective adsorption of the solubilized proteins has been achieved using immobilized biospecific ligands, shown in Table 11.1. By this way, for example, a number of authors succeeded in the purification of hormone receptors (for reviews, see Cuatrecasas and Hollenberg, 1976; Cooke, 1982) that have affinity constants for ligand binding in the range of 10^{-9}-10^{-10} M. In general, these high affinities are not dramatically altered upon detergent solubilization. Affinity chromatography allowed in some cases a more than 1000-fold purification in one-step procedures.

Also widely applied was the lectin affinity chromatography (Lotan and Nicolson, 1979), first to surface membrane proteins, which always seem to be glycosylated (Gahmberg, 1976). This method is discussed in Chapter 10.

There are only few examples for the application of hydrophobic chromatography (Liljas et al., 1976; Simonds and Yon, 1977). Apparently, this method provides little selectivity for membrane constituents because all the detergent-solubilized proteins have large hydrophobic regions.

At least in some cases the hydrophobic interactions limit the specificity of the adsorption step. This can be partially overcome by high detergent concentrations and a low polarity of the buffer medium (Schunck et al., 1983). However, hydrophobic chromatography can be used advantageously to isolate the membrane-binding domains after proteolytic cleavage of the purified proteins (Tsuchida and Sato, 1983).

The highly purified preparations of integral membrane proteins obtained by affinity chromatography provide a key for future studies in such important fields as the mechanism of hormone action, substrate translocation, receptor-mediated endocytosis, electron transfer processes, and cell-cell interactions.

REFERENCES

Akhrem, A. A., Lapko, V. N., Lapko, A. G., Shkumatov, V. M., and Chashchin, V. L. (1979). Isolation, structural organization and mechanism of action of mitochondrial steroid hydroxylating systems. *Acta Biol. Med. Germ. 38*: 257-273.

Ashwell, G., and Harford, J. (1982). Carbohydrate-specific receptors of the liver. *Annu. Rev. Biochem. 51*: 531-554.

Azzi, A., Brodbeck, U., and Zahler, P. (1981). *Membrane Proteins — A Laboratory Manual*. Springer-Verlag, Berlin.

Barbero, M. C., Rial, E., Otamendi, J. J., Gurtubay, J. I. G., and Goni, F. M. (1982). Fractionation of rat liver mitochondrial components after short treatments with Triton X-100. *Int. J. Biochem. 14*: 933-940.

Bendzko, P., and Pfeil, W. (1983). Thermodynamic investigations of cytochrome b_5 unfolding. II. Detergent solubilized cytochrome b_5 in solution and in a reconstituted system with dimyristoyl phosphatidylcholine. *Biochim. Biophys. Acta 742*: 669-676.

Bill, K., Casey, R. P., Broger, C., and Azzi, A. (1980). Affinity chromatography purification of cytochrome c oxidase; use of a yeast cytochrome c-thiol-Sepharose 4 B column. *FEBS Lett. 120*: 248-250.

Blobel, G. (1980). Intracellular protein topogenesis. *Proc. Nat. Adad. Sci. U.S. 77*: 1496-1500.

Brock, C. J., and Tanner, M. J. A. (1982). Structure and synthesis of integral membrane proteins. In *Biological Membranes*, Vol. 4, D. Chapman (ed.). Academic Press, London, pp. 75-130.

Brodbeck, U., Gentinetta, R., and Ott, P. (1981). Purification by affinity chromatography of red cell membrane acetylcholine esterase. In *Membrane Proteins — A Laboratory Manual*, A. Azzi, U. Brodbeck, and P. Zahler (eds.). Springer-Verlag, Berlin, pp. 85-96.

Capaldi, R. A. (1977). The structural properties of membrane proteins. In *Membrane Proteins*, R. A. Capaldi (ed.). Marcel Dekker, New York, pp. 1-20.

Capaldi, R. A. (1982). Structure of intrinsic membrane proteins. *TIBS 7*: 292-295.

Caroni, P., and Carafoli, E. (1981). The Ca^{2+}-pumping ATPase of heart sarcolemma. Characterization, calmodulin dependence, and partial purification. *J. Biol. Chem. 256*: 3263-3270.

Cerione, R. A., Strulovici, B., Benovic, J. L., Lefkowitz, R. J., and Caron, M. G. (1983). Pure β-adrenergic receptor: The single polypeptide confers catecholamine responsiveness to adenylate cyclase. *Nature 306*: 562-566.

Chapman, D. (1982). Protein-lipid interactions in model and natural biomembranes. In *Biological Membranes*, Vol. 4, D. Chapman (ed.). Academic Press, London, pp. 179-230.

Clarke, S. (1975). The size and detergent binding of membrane proteins. *J. Biol. Chem. 250*: 5459-5469.

Coleman, R. (1973). Membrane-bound enzymes and membrane ultrastructure. *Biochim. Biophys. Acta 300*: 1-30.

Conti-Tronconi, B. M., and Raftery, M. (1982). The nicotinic cholinergic receptor: Correlation of molecular structure with functional properties. *Annu. Rev. Biochem. 51*: 491-530.

Cooke, B. A. (1982). Plasma membrane receptors and regulation of adenylate cyclase. In *Biological Membranes*, Vol. 4, D. Chapman (ed.). Academic Press, London, pp. 327-366.

Cuatrecasas, P., and Hollenberg, M. D. (1976). Membrane receptors and hormone action. *Advan. Protein Chem. 30*: 252-451.

Cuatrecasas, P., and Parikh, J. (1974). Insulin receptors. *Methods Enzymol. 34*: 653-670.

Dratz, E. A., and Hargrave, P. A. (1983). The structure of rhodopsin and the rod outer segment disk membrane. *TIBS 8*: 128-131.

Engelman, D. M., Henderson, R., McLachlan, A. D., and Wallace, B. A. (1980). Path of the polypeptide in bacteriorhodopsin. *Proc. Nat. Acad. Sci. U.S. 77*: 2023-2027.

Eyton, G. D. (1982). Use of liposomes for reconstitution of biological functions. *Biochim. Bipohys. Acta 694*: 185-202.

Finean, J. B., and Michell, R. H. (1981). Membrane structure. *New Comp. Biochem. 1*: 1-36.

Forsee, W. T., and Schutzbach, J. S. (1981). Purification and characterization of a phospholipid-dependent α-mannosidase from rabbit liver. *J. Biol. Chem. 256*: 6577-6582.

Freedman, R. B. (1981). Membrane-bound enzymes. *New Comp. Biochem. 1*: 161-214.

Fujita-Yamaguchi, Y., Choi, S., Sakamoto, Y., and Itakura, K. (1983). Purification of insulin receptor with full binding activity. *J. Biol. Chem. 258*: 5045-5049.

Fuller, S. D., Capaldi, R. A., and Henderson, R. (1979). Structure of cytochrome c oxidase in deoxycholate-derived two-dimensional crystals. *J. Mol. Biol. 134*: 305-327.

Furth, A. J. (1980). Removing unbound detergent from hydrophobic proteins. *Anal. Biochem. 109*: 207-215.

Gahmberg, G. C. (1976). External labeling of human erythrocyte glycoproteins; studies with galactose oxidase and fluorography. *J. Biol. Chem. 251*: 510-515.

Gautheron, D. C., and Julliard, J. H. (1979). Isolation of a glutamate carrier system from pig heart mitochondria and incorporation into liposomes. *Methods Enzymol. 56*: 419-430.

Geiger, B. (1983). Membrane-cytoskeleton interactions. *Biochim. Biophys. Acta 737*: 305-341.

Gennis, R. B., and Jonas, A. (1977). Protein-lipid interactions. *Annu. Rev. Biophys. Bioeng. 6*: 195-238.

Graham, R. M., Hess, H.-J., and Homcy, C. J. (1982). Solubilization and purification of the α_1-adrenergic receptor using a novel affinity resin. *Proc. Nat. Acad. Sci. U.S. 79*: 2186-2190.

Guidotti, G. (1972). Membrane proteins. *Annu. Rev. Biochem. 41*: 731-752.

Heidman, T., and Changeux, J. P. (1978). Structural and functional properties of the acetylcholine receptor protein in its purified and membrane-bound states. *Annu. Rev. Biochem. 47*: 317-357.

von Heijne, G. (1981). Membrane proteins; the amino acid composition of membrane penetrating segments. *Eur. J. Biochem. 120*: 275-278.

Heinemann, F. S., and Ozols, J. (1983). The complete amino acid sequence of rabbit phenobarbital-induced liver microsomal cytochrome P-450. *J. Biol. Chem. 258*: 4195-4201.

Helenius, A., McCaslin, D. R., Fries, E., and Tanford, C. (1979). Properties of detergents. *Methods Enzymol. 56*: 734-749.

Helenius, A., and Simons, K. (1972). The binding of detergents to lipophilic and hydrophilic proteins. *J. Biol. Chem. 247*: 3656-3661.

Helenius, A., and Simons, K. (1975). Solubilization of membranes by detergents. *Biochim. Biophys. Acta 415*: 29-79.

Imai, Y., and Sato, R. (1974). An affinity column method for partial purification of cytochrome P-450 from phenobarbital-induced rabbit liver microsomes. *J. Biochem. 75*: 689-697.

Israelachvili, J. N., Marcelja, S., and Horn, R. G. (1980). Physical principles of membrane organization. *Q. Rev. Biophys. 13*: 121-200.

Kawasaki, T. Etoh, R., and Yamashina, J. (1978). Isolation and characterization of a mannan binding protein from rabbit liver. *Biochem. Biophys. Res. Commun. 81*: 1018-1024.

Kracke, G. R., O'Neal, S. G., and Chacko, G. K. (1981). Partial purification and characterization of $(Na^{+}+K^{+})$-ATPase from Garfish olfactory nerve axon plasma membrane. *J. Membr. Biol. 63*: 147-156.

Kreil, G. (1981). Transfer of proteins across membranes. *Annu. Rev. Biochem. 50*: 317-348.

Laemmli, U. K. (1970). Cleavage of structural proteins during the assembly of the head of bacteriophage T4. *Nature 227*: 680-685.

Lees, M. B., Sakura, J. D., Sapirstein, V. S., and Curatolo, W. (1979). Structure and function of proteolipids in myelin and non-myelin membranes. *Biochim. Biophys. Acta 559*: 209-230.

Lichtenberg, D., Robson, R. J., and Dennis, E. D. (1983). Solubilization of phospholipids by detergents; structural and kinetic aspects. *Biochim. Biophys. Acta 737*: 285-304.

Liljas, L., Lundahl, P., and Hjerten, S. (1976). The major sialoglycoprotein of the human erythrocyte membrane; release with a nonionic detergent and purification. *Biochim. Biophys. Acta 426*: 526-534.

Ling, K. Y. L., and Faust, R. G. (1983). Reconstitution of a partially purified Na^+-independent D-glucose transport system from rat jejunal basalolateral membranes. *Int. J. Biochem. 15*: 27-34.

Lotan, R., and Nicolson, G. L. (1979). Purification of cell membrane glycoproteins by lectin affinity chromatography. *Biochim. Biophys. Acta 559*: 329-376.

MacDonald, R. J. (1980). Action of detergents on membranes: Differences between lipid extraction from red cell ghosts and from red cell lipid vesicles by Triton X-100. *Biochemistry 19*: 1916-1922.

Makino, S., Reynolds, J. A., and Tanford, C. (1973). The binding of deoxycholate and Triton X-100 to proteins. *J. Biol. Chem. 248*: 4926-4932.

Moore, P. B., and Kraus-Friedmann, N. K. (1983). Hepatic microsomal Ca^{2+}-dependent ATPase; calmodulin dependence and partial purification. *Biochem. J. 214*: 69-75.

Newman, M. J., Foster, D. L., Wilson, T. H., and Kuback, H. R. (1981). Purification and reconstitution of functional lactose carrier from E. coli. *J. Biol. Chem. 256*: 11804-11808.

Nicolson, G. L. (1976). Transmembrane control of the receptors on normal and tumor cells. I. Cytoplasmic influence over cell surface components. *Biochim. Biophys. Acta 457*: 57-108.

Nozaki, Y., Reynolds, J. A., and Tanford, S. (1974). The interaction of a cationic detergent with bovine serum albumin and other proteins. *J. Biol. Chem. 249*: 4452-4459.

O'Farell, P. H. (1975). High resolution two-dimensional electrophoresis of proteins. *J. Biol. Chem. 250*: 4007-4013.

Pfeiffer, F., Graham, D., and Betz, H. (1982). Purification by affinity chromatography of the glycine receptors of rat spinal cord. *J. Biol. Chem. 257*: 9389-9393.

Prado, A., Arrondoi, J. L. R., Villena, A., Goni, F. M., and Macarulia, J. M. (1983). Membrane-surfactant interactions; the effect of Triton X-100 on sarcoplasmic reticulum vesicles. *Biochim. Biophys. Acta 733*: 163-171.

Pricer, W. E., Hudgin, R. K., Ashwell, G., Stockert, R. J., and Morell, A. G. (1974). A membrane receptor protein for asialoglycoprotein. *Methods Enzymol. 34*: 688-695.

Racker, E. (1979). Reconstitution of membrane processes. *Methods Enzymol. 55*: 699-711.

Rascati, R. J., and Parsons, P. (1979). Purification and characterization of cytochrome c oxidase from rat liver mitochondria. *J. Biol. Chem. 254*: 1586-1593.

Reynolds, J. A., and Tanford, C. (1970). The gross conformation of protein-sodium dodecyl sulfate complexes. *J. Biol. Chem. 245*: 5161-5165.

Riege, P., Schunck, W.-H., Honeck, H., and Müller, H.-G. (1981). Cytochrome P-450 from Lodderomyces elongisporus: Its purification and some properties of the highly purified protein. *Biochem. Biophys. Res. Commun. 98*: 527-534.

Rizzolo, L. J. (1981). Kinetics and protein subunit interactions of E. coli phosphatidylserine decarboxylase in detergent solutions. *Bio chemistry 20*: 868-873.

Robinson, N. C., and Tanford, C. (1975). The binding of deoxycholate, Triton X-100, sodium dodecyl sulfate and phosphatidylcholine vesicles to cytochrome b_5. *Biochemistry 14*: 369-377.

Sato, R., and Omura, T. (1979). *Cytochrome P-450*. Kodansha-Academic Press, Tokyo.

Schlesinger, M. J. (1981). Proteolipids. *Annu. Rev. Biochem. 50*: 193-206.

Schmidt, U. M., Eddy, B., Fraser, C. M., Venter, J. C., and Semenza, G. (1983). Isolation of (a subunit of) the Na^+/D-glucose cotransporter(s) of rabbit intestinal brush border membranes using monoclonal antibodies. *FEBS Lett. 161*: 279-283.

Schneider, R. J., Kulczycki, A., Jr., Law, S. K., and Atkinson, J. P. (1981). Isolation of a biologically active macrophage receptor for the third component of complement. *Nature 290*: 789-792.

Schunck, W.-H., Riege, P., Honeck, H., and Müller, H.-G. (1983). Purification and molecular properties of alkane-induced cytochrome P-450 from the yeast Lodderomyces elongisporus. *Biochimia 48*: 518-526.

Shepard, E. A., Pike, S. F., Rabin, B. R., and Phillips, J. R. (1983). A rapid one-step purification of NADPH-cytochrome c (P-450) reductase from rat liver microsomes. *Anal. Biochem. 129*: 430-433.

Shiu, R. P. C., and Friesen, H. G. (1974). Solubilization and purification of a prolactin receptor from the rabbit mammary gland. *J. Biol. Chem. 249*: 7902-7911.

Sigel, E., Stephenson, A., Mamalaki, C., and Barnard, E. (1983). A γ-aminobutyric acid/benzodiazepine receptor complex of bovine cerebral cortex. *J. Biol. Chem. 258*: 6965-6971.

Simmonds, R. J., and Yon, R. J. (1977). Protein chromatography on adsorbents with hydrophobic and ionic groups; purification of human erythrocyte glycophorin. *Biochem. J. 163*: 397-400.

Simons, K., Garoff, H., and Helenius, H. (1977). The glycoproteins of the Semliki Forest virus membranes. In *Membrane Proteins*, R. A. Capaldi (ed.). Marcel Dekker, New York, pp. 207-234.

Singer, S. J. (1974). The molecular organization of membranes. *Annu. Rev. Biochem. 43*: 805-833.

Singer, S. J. (1977). Thermodynamics, the structure of integral membrane proteins, and transport. *J. Supramol. Struct. 6*: 313-323.

Singer, S. J., and Nicolson, G. L. (1972). The fluid mosaic model of the structure of cell membranes. *Science 175*: 720-731.

Spatz, L., and Strittmatter, P. (1971). A form of cytochrome b_5 that contains an additional hydrophobic sequence of 40 amino acid residues. *Proc. Nat. Acad. Sci. U.S. 68*: 1042-1046.

Szoka, F., and Papahadjopoulos, D. (1980). Comparative properties and methods of preparation of lipid vesicles (liposomes). *Annu. Rev. Biophys. Bioeng. 9*: 476-508.

Tanford, C. (1980). *The Hydrophobic Effect: Formation of Micelles and Biological Membranes*. John Wiley & Sons, New York.

Tanford, C., and Reynolds, J. A. (1976). Characterization of membrane proteins in detergent solutions. *Biochim. Biophys. Acta 457*: 133-170.

Thompson, D. A., and Ferguson-Miller, S. (1983). Lipid and subunit III depleted cytochrome c oxidase purified by horse cytochrome c affinity chromatography in lauryl maltoside. *Biochemistry 22*: 3178-3183.

Tomita, M., and Marchesi, V. T. (1975). Amino acid sequence and oligosaccharide attachment sites of human erythrocyte glycophorin. *Proc. Nat. Acad. Sci. U.S. 72*: 2964-2968.

Tsuchida, S., and Sato, K. (1983). Purification of detergent-solubilized form and membrane-binding domain of rat γ-glutamyltransferase by immuno-affinity and hydrophobic chromatography. *Biochim. Biophys. Acta 756*: 341-348.

Uterman, G., and Simons, K. (1974). Studies on the amphiphatic nature of the membrane proteins in Semliki Forest virus. *J. Mol. Biol. 85*: 569-587.

Warren, G. (1981). Membrane proteins: Structure and assembly. *New Comp. Biochem. 1*: 215-251.

Waters, M. J., and Friesen, H. G. (1978). Purification of a growth hormone receptor, and its use to generate specific antibodies. *Proc. Aust. Biochem. Soc. 11*: 101.

Weber, K., Pringle, J. R., and Osborn, M. (1969). Measurements of molecular weights by electrophoresis on SDS-acrylamide gel. *Methods Enzymol. 26*: 3-27.

Yasukochi, Y., and Masters, B. S. S. (1976). Some properties of a detergent-solubilized NADPH-cytochrome c (cytochrome P-450) reductase purified by biospecific affinity chromatography. *J. Biol. Chem. 251*: 5337-5344.

Zahler, P., Wüthrich, C., and Wolf, M. (1981). Isolation of a membrane protein: The phospholipase A_2 from sheep erythrocyte membranes. In *Membrane Proteins — A Laboratory Manual*, A. Azzi, U. Brodbeck, and P. Zahler (eds.). Springer-Verlag, Berlin, pp. 66-69.

part IV
VARIANTS OF AFFINITY CHROMATOGRAPHY OTHER THAN BIOADSORPTION

12 Dye-Ligand Chromatography

Reactive dyes, a class of synthetic textile dyes, are successfully used as immobilized ligands in affinity chromatography. Since the introduction of Cibacron Blue F3G-A* for the purification of nucleotide cofactor-dependent enzymes, this field has been greatly expanded and other dyes investigated as selective ligands. The reactive dyes consist of the dye component (anthraquinone, azo, pyrazolone dyes) and special reactive residues through which they are covalently bound to the fibrous material, such as chlorotriazinyl, vinyl sulfone, epoxide, or ethyleneimine residues, as well as sulfuric acid esters of β-hydroxyethylsulfones. Since the reactive dyes have no biological relation to the enzymes to be purified, the term "pseudo-affinity chromatography" was introduced, in contradistinction to the genuine biospecific affinity chromatography (Haff and Easterday, 1978).

In 1968, the interaction between the dextran blue (Pharmacia, Uppsala, Sweden), a colored derivative of a dextran fraction with an average molecular weight of 2,000,000 and Cibacron Blue F3G-A as chromophores, and pyruvate kinase (Haeckel et al., 1978) or yeast phosphofructokinase (Kopperschläger et al., 1968) was discovered, providing a basis for the development of affinity chromatographic procedures. Shortly afterward the principle was expanded to other enzymes and proteins, and it was shown that the blue dye is responsible

---

*This dye is also known as Cibacron Blue 3G-A (now redesignated) and Procion Blue H-B, color index 61211

for the binding. Therefore, Cibacron Blue F3G-A coupling directly to Sephadex was used as affinity medium for the purification of yeast pyruvate kinase (Röschlau and Hess, 1972) and yeast phosphofructokinase (Böhme et al., 1972).

This reactive dye became established as a group-selective ligand either in direct form or as the high-molecular-weight derivatives dextran blue. This "Blue Dextran" has been also used as an entrapped ligand in polyacrylamide gel (Kopperschläger et al., 1971) or in soluble form (Haeckel et al., 1968).

Since Cibacron Blue F3G-A is the most important representative, the basic principles of dye-ligand chromatography is explained here by this example.

STRUCTURE, PROPERTIES, AND IMMOBILIZATION OF CIBACRON BLUE F3G-A

The whole structure of Cibacron Blue F3G-A may be divided into four parts (Fig. 12.1):

1. The sulfonated anthraquinone group (a)
2. The bridging sulfonated benzene ring (b)
3. The triazine ring (c)
4. The terminal sulfonated benzene ring (d)

It possesses both a hydrophobic and an anionic character and is able to form hydrogen bridges with complementary structures via amino or carbonyl groups. The sulfonic acid residues facilitate solubility in water.

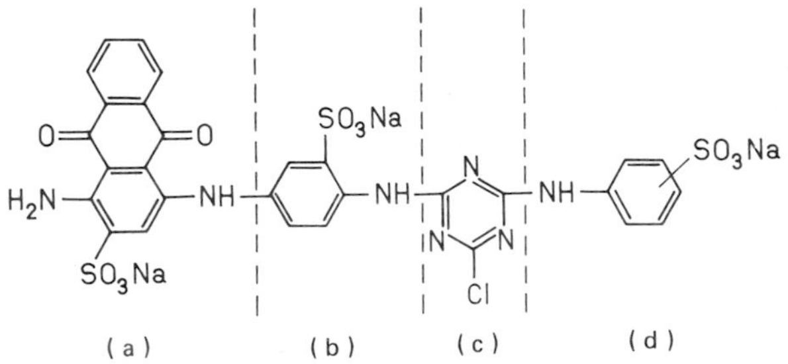

FIGURE 12.1 Four structural parts of Cibacron Blue F3G-A. (a) Sulfonated anthraquinone group. (b) Bridging sulfonated benzene ring. (c) Triazine ring. (d) Terminal sulfonated benzene ring.

The commercial dye is pure to about 60-80%. Thin-layer chromatographic analysis has shown that the dye contains several minor chromophores, which neither reacted with the support nor influenced the interaction with the protein (Weber et al., 1979; Amicon, 1980; Birkenmeier, 1981). At present, cross-linked agarose dominates as the support material. In addition, other supports are used, including cross-linked dextran, cellulose, polyacrylamide, and silica. The choice of the support may be of decisive importance for effectivity (Angal and Dean, 1977). The covalent binding of the dye can be achieved in different ways. Depending on the type and site of covalent linkage, however, the properties of the affine support will be different (see below).

By analogy with the methods used in the textile industry, the dye may be covalently bound to polysaccharides in sodium chloride-sodium carbonate solution at 40-80°C (Böhme et al., 1972). The binding is accomplished by nucleophilic exchange of the chlorine atom at the triazine ring with formation of an ether bond (Chap, 4, Polysaccharides, Other Methods).

A modified procedure was described for coupling to bead cellulose and its hydroxyethyl derivatives (Pommerening et al., 1982). The coupling can be carried out very successfully in a mixture of acetone-water (70:30 v/v) containing 70 μmol sodium hydroxide per gram moist swollen bead cellulose. The yield of covalently bound dye is higher in comparison with the coupling in water-sodium carbonate. The reaction is promoted by OH^-, but a great excess of OH^- leads to partial degradation of the dye. The OH groups of hydroxyethyl side chains are more reactive than OH groups of the glucose units by both methods. This is attributed partly to steric hindrance at the coupling procedure.

Another way is the direct covalent attachment of the dye to the matrix by the CNBr method (Röschlau and Hess, 1972) or periodate oxidation of the support (Dean and Watson, 1979). In both cases binding is accomplished via the NH_2 group at the anthraquinone ring.

Coupling of Blue Dextran by the CNBr method is also achieved via the primary amino group of the anthraquinone ring. As a result, however, the dye is not uniform (Fig. 12.2). Part of it is bound to the dextran; another part realizes additional binding to the matrix via the primary amino group.

The load density may be varied in broad limits. With agarose as support, a load density of approximately 2 μmol/ml or 2.5-4.0 mg/ml swollen gel has proved optimal. By variation of experimental conditions, for example, Sephadex G100-Cibacron Blue F3G-A conjugates have been prepared with the dye content varying between 6 and 840 mg dye per gram Sephadex G100 (Kopperschläger et al., 1982).

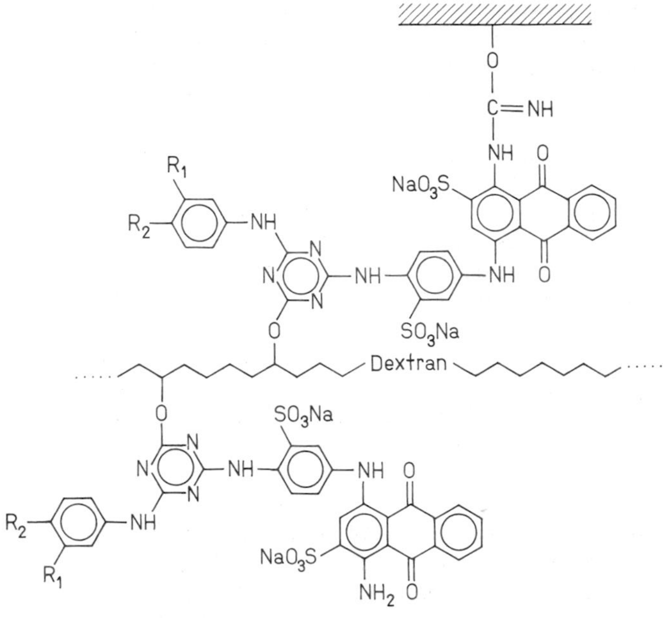

FIGURE 12.2 Partial structure of Blue Dextran-Sepharose.

MECHANISM OF DYE-PROTEIN INTERACTION

Cibacron Blue F3G-A has affinity for a variety of enzymes and other proteins. Surveys with numerous literature references have been published repeatedly (Dean and Watson, 1979; Amicon, 1980; Kopperschläger et al., 1982). More than 80 enzymes and other biomacromolecules can be purified by means of this dye: the largest two classes are dehydrogenases and kinases. Other enzymes or proteins include transferases, cyclases, reductases, serum proteins, and interferon. To date it is not clear in detail why Cibacron Blue F3G-A and other dyes can be used as group-specific ligands. Extensive investigations have been carried out to elucidate the mechanism, including a direct study of the interaction of protein with the dye or its analogs. The empirical data of purification suggest that different binding mechanisms may exist for the different enzymes and proteins, attributed to the chemical and steric structure of this dye. The largest group of proteins that binds to the Cibacron Blue F3G-A are nucleotide cofactor-dependent enzymes. By virtue of its flexibility, Cibacron Blue F3G-A should be able to mimic the conformation of NAD^+.

Thompson and colleagues (1975) have demonstrated the coincident spatial arrangement of the two molecules. Stellwagen (1977) and co-workers have developed a theoretical concept for the interaction between Cibacron Blue F3G-A and enzymes. Their hypothesis suggests that the dye interacts specifically with a particular feature of most cofactor-dependent enzymes or other proteins called "dinucleotide fold." The dinucleotide folds are identical with the binding sites of cofactors, such as NAD^+, $NADP^+$, and ATP. This structure is a supersecondary structure, found in many dehydrogenases, kinases, and other enzymes and preserved during evolution. The enzymes investigated by them show not only a strong correlation between dye interaction and the presence of a dinucleotide fold, but also an analogous elution behavior. All of them can be eluted by biospecific elution with a low concentration of a suitable cofactor (0.1-1.0 mM) or a high concentration of NaCl (greater than 100 mM). Biomacromolecules that do not have a dinucleotide fold but are bound to the Blue gel can be eluted only with a cofactor greater than 10 mM or with less than 100 mM NaCl (Thompson et al., 1975).

Other results are not in agreement with this concept.

Several proteins and enzymes that do not possess a dinucleotide fold, such as serum albumin, interferon, and phosphodiesterase, are adsorbed very strongly and specifically to matrix-bound Cibacron Blue F3G-A and may be eluted only with a high concentration of salts or a low concentration of specific substrates.

Matthes (1979) shows the existence of similar cleft-shaped binding sites in proteins that do not have the dinucleotide fold.

The arabinose binding protein isolated from *Escherichia coli* has a dinucleotide fold, but shows no specific binding (Beissner et al., 1979).

Subramanian and Kaufman (1980) proposed on the basis of results with dihydrofolate reductase that the interaction takes place with proteins possessing either a cluster of an apolaric residue or of positively charged groups, which bind to the aromatic rings or sulfonate residues of the dye.

Several authors (Bessner et al., 1979), therefore, concluded that the triazine dyes are only cation exchangers with hydrophobic properties, not strictly bioaffine ligands. This conclusion goes too far. Beyond the examples already mentioned there is sufficient circumstantial evidence that, in many cases, the dye molecule interacts very specifically with ligand binding sites of enzymes. Furthermore, steric factors have an influence. Cibacron Brilliant Blue FBR-P differs from Cibacron Blue F3G-A only in the position of sulfonic acid and the phenylenediamine ring (Fig. 12.3) but does not show any affinity to

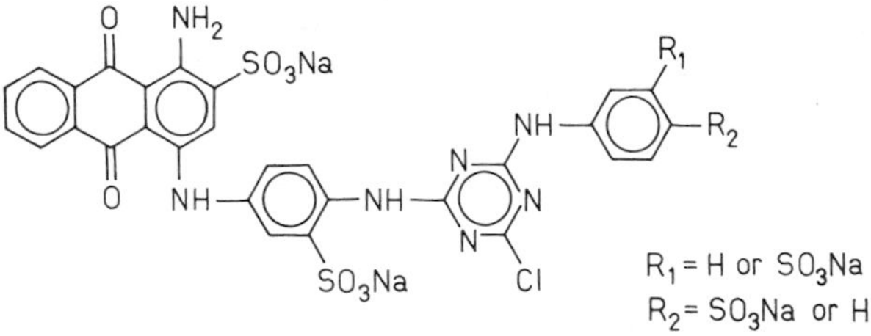

R_1 = H or SO_3Na
R_2 = SO_3Na or H

Cibacron Blue F3G-A

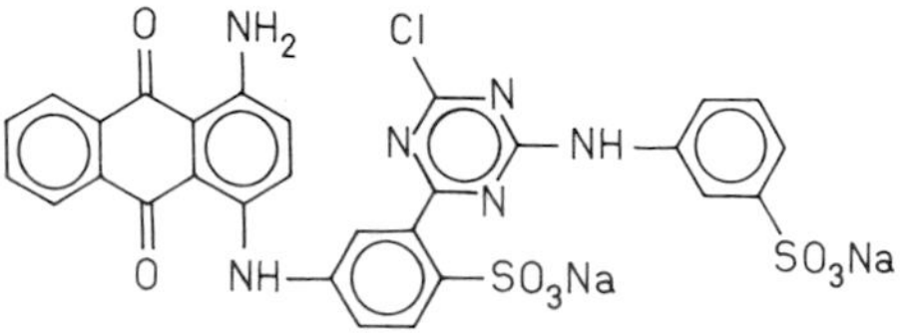

Cibacron Brilliant Blue FBR-P

FIGURE 12.3 Chemical structure of Cibacron Blue F3G-A and Cibacron Brilliant Blue FBR-P.

phosphofructokinase (Böhme et al., 1972). An analogous behavior of the two dyes has been found with the interaction of human serum. Cibacron Blue F3G-A adsorbs albumin specifically. The other dye also binds this serum protein (Travis et al., 1976), but this interaction is unspecific. It is not possible to separate albumin from the whole serum with this dye (Pommerening et al., 1982) (see also Purification of Blood Proteins). In many cases the effectiveness of dye-ligand chromatography is influenced by the nature of its binding to the matrix. Sometimes it may be advantageous to use Dextran Blue instead of the pure dye, as for the purification of interferon from human or mouse sources, tRNA synthetase, or human clotting factor X. On the other hand, Blue Dextran agarose was less effective than agarose with directly coupled Cibacron Blue F3G-A for the purification of interferon from hamster or horse, for example. In several cases the enzymes showed no interaction with Blue Dextran, but high affinity to the pure dye. The reason possibly could be steric factors due to dextran chain. A detailed discussion was given in the paper by Amicon (1980).

The interaction is also dependent on the type of dye binding to the matrix. In some cases the anthraquinone group is essential for the adsorption of the protein. By coupling the dye via the primary amino group in position 1 of the anthraquinone ring, this part of the ligand is sterically blocked. It was found that the binding of proteins is then either weakened or eliminated, as in the adsorption of interferon (Bollin et al., 1978) or albumin (Travis et al., 1976).

The concentration of the dye has an important influence on binding strength and capacity. Details are discussed of the interaction with serum proteins (see Purification of Blood Proteins). Angal and Dean (1977) found an influence of the matrix on the binding of albumin to immobilized Cibacron Blue F3G-A. For example, the dye-cellulose has a low albumin binding capacity (1/10 to 1/100 of the capacity of Blue-agarose). The accessibility of the dye-ligand with cellulose and several dextran derivatives appears to be less than that with agarose as matrix. Probably this is due to differences in the molecular conformation of the polysaccharide chains. The albumin binding capacity of blue-cellulose can be drastically increased with macroporous bead cellulose as matrix (Pommerening et al., 1982). This is likely to be due to the different accessibility of the ligand in the two matrices. This shows that not only the chemical structure but also the physical structure can be responsible for the interaction between dye and proteins.

This brief outline shows that a variety of factors are generating or influencing the interactions. This is why the mechanism of the binding of dyes to the proteins is not simple and has not been explained fully. Nevertheless, several conclusions can be derived. The interaction is of ionic and/or hydrophobic (apolaric) nature. Both properties have a varying share in the binding, which differs from protein to protein and from dye to dye. Partial structures of the dye are essential for the binding of many proteins. The interaction correlates with the presence of certain protein structures. In several cases the dye binds to the ligand binding sites with high specificity.

PURIFICATION OF ENZYMES

Cibacron Blue F3G-A is a very suitable group-specific ligand for enzymes requiring adenylyl-containing cofactors, particularly for dehydrogenases and kinases. The other enzymes are not as simple to classify (CoA enzymes, hydrolases, transferases, nucleases, polymerases, synthetases, and others). Recently, experimental conditions for the purification of these enzymes were summarized by Dean and Watson (1979) and Amicon (1980), as well as Kopperschläger and colleagues (1982).

Cibacron Blue F3G-A can bind both NAD^+ – and $NADP^+$ – as well as ATP-dependent enzymes from many species and tissues; for example, 22 lactate dehydrogenases and 14 different kinases have been shown to bind to this dye. Likewise it is possible to separate isoenzymes, for instance, from malate dehydrogenase (Hägele et al., 1978), lactate dehydrogenase (Gordon and Doelle, 1976), or phosphofructokinase (Cottreau et al., 1979).

For desorption the oxidized or reduced cofactors can be used, mostly with low concentrations (<10 mM). An unspecific elution can also be carried out by increasing salt concentrations (0.1-2.0 M).

In the last few years very interesting purification procedures of enzymes have been published for genetic engineering: restriction endonucleases (Baksi and Rushizky, 1978, 1979a, b; Baksi et al., 1978), tRNA synthetases (Nikodem et al., 1977), and T_4-DNA ligase (Sugiura, 1980).

PURIFICATION OF BLOOD PROTEINS

Blood plasma is a complex mixture containing more than 100 different proteins. Often the isolation of several proteins is complicated by the fact that more than 50% of the protein content consists of albumin and that this albumin has physical properties similar to other proteins with very low plasma levels. Therefore, it is of practical interest to develop effective methods for the isolation and purification of albumin as well as for the preparation of albumin-depleted plasma. To this end, affinity chromatography on Cibacron Blue F3G-A is a very effective method. Travis and Pannell (1973) were the first to show that albumin can be removed from other proteins by chromatography with Dextran Blue-Sepharose. The direct linkage of the dye to the agarose yields a more effective gel for the selective removal of albumin (Travis et al., 1976). The process is mild, and the matrix has a high capacity. Under normal conditions (0.02-0.05 M buffer, pH 7-8.5, containing 0.0-0.5 M NaCl) more than 95% of the albumin is removed with little nonspecific adsorption of other proteins. The elution of albumin proceeds predominantly with 1-3 M NaCl or 0.2-2 M NaSCN. Other eluents are KCl, N-acetyltryptophan, caprylic acid, $CaCl_2$, and 50% ethanol, pH 2.4 (Harvey, 1980). In addition to pH, temperature, and ionic strength, the nature of the matrix has an effect on the binding of albumin to the immobilized dye (Angal and Dean, 1977). These authors and Birkenmeier (1981) found that normal cellulose is not suitable as matrix for the separation of the albumin. In the case of macroporous bead cellulose, however, this dye-matrix is very suitable for the selective adsorption of human serum albumin (Pommerening et al., 1982).

Table 12.1 shows that the adsorption of other proteins increases with the dye concentration and that the interaction of Ostazin Brilliant Blue (another trade name for Cibacron Brilliant Blue FBR-P) with albumin is not specific. This demonstrates among other things the importance of steric factors for the interaction between albumin

TABLE 12.1 Recovery of Serum Proteins After Incubation with Dye-Gels[a]

Blue-gel	Dye[b]	Serum fraction (%)					
		TP[c]	HSA[d]	α_1	α_2	β	γ
Blue-Sepharose	2.0[e]	56.5	38.6	70.4	84.6	90.1	75.0
Blue BC[f]	2.9	41.4	7.1	93.3	83.3	83.3	90.9
Blue-BC	4.4	30.3	4.7	60.0	86.1	61.1	72.7
Blue-BC	7.3	29.1	3.9	85.2	73.8	64.8	52.3
Os-Blue-BC[g]	8.9[h]	65.6	59.2	51.9	70.8	73.2	86.4

[a]Adsorption conditions: 10 ml Blue-gel, 2 ml human serum, 0.05 M tris·HCl-0.5 M NaCl, 20-24°C, column or batch technique (4-5 hr).
[b]μmole dye per g swollen moist bead cellulose.
[c]Total proteins.
[d]Human serum albumin.
[e]μmol/ml.
[f]Cibacron Blue F3G-A bead cellulose.
[g]Ostazin Brilliant Blue HBR-bead cellulose.
[h]mg dye per g moist swollen bead cellulose.

and the dye. Albumins from animal plasma bind to Blue-Sepharose with different affinities (Kelleher et al., 1979; Mahany et al., 1981; Naval et al., 1982). The apparent affinity constants differ between $3.9 \times 10^4\ M^{-1}$ and $0.9 \times 10^4\ M^{-1}$ in the order human, rabbit, horse, pig, dog, cow, rat, and chicken. Beyond albumin, other plasma proteins, too, show interaction with the dye. In addition to the high affinity of albumin there are proteins of mean (immunoglobulins, transferrin, and haptoglobins) and weak affinity (α_2-macroglobulin, prealbumin, α_1-proteinase inhibitor, and α_1-acidic glycoprotein). With the increase in H^+ concentration and/or dye concentration the adsorption of serum proteins increases and may lead to complete adsorption. Frequently this different affinity has been used for the isolation or purification of the plasma proteins. Some examples are summarized in Table 12.2

Gianazza and Arnaud (1982) found that the chromatography of plasma on immobilized Cibacron Blue F3G-A can be a useful step in the purification of plasma proteins. They investigated the behavior of 27 different proteins. Of these, 4 proteins (α_1-antitrypsin,

TABLE 12.2 Some Purification of Plasma Proteins Without Albumin by Chromatography on Immobilized Cibacron Blue F3G-A

Protein	Reference
Lipoproteins	Wille (1976), Leatherbarrow and Dean (1980), Saint-Blancard et al. (1982)
α-Fetoprotein	Kelleher et al. (1979), Birkenmeier et al. (1983).
α_2-Macroglobulin	Virca et al. (1978), Arnaud and Gianazza (1982), Bridges et al. (1982).
IgGs	Saint-Blancard et al. (1982)
Human blood clotting factors II and IX	Swart and Henker (1970)
Blood clotting factor X	Vician and Tishkoff (1976)
Complement C_1-C_9	Gee et al. (1979)
α_1-Antichymotrypsin	Travis et al. (1978)
α_1-Proteinase inhibitor	Pannell et al. (1974), Birkenmeier (1981)
α_1-Acid glycoprotein	Birkenmeier (1981)
α_1-Antitrypsin, ceruplasmin, antithrombin III, hemopexin	Gianazza and Arnaud (1982)

ceruloplasmin, antithrombin III, and hemopexin) were enriched between 10- and 75-fold, and the recovery of the proteins ranged from 52 to 95%. Recently, a new method has been described for the purification of IgGs and albumin on a large scale by ion-exchange and affinity chromatography with Blue Trisacryl (Saint-Blancard et al., 1982). The affine matrix was obtained by coupling Cibacron-Blue F3G-A on a chemically modified macroporous and nonionic Trisacryl.

The starting material was 17 liters human plasma cryosupernatant or fraction I supernatant, and only three chromatographic steps at 20°C were needed, each with a column volume of 50 liters. The elution profile is given in Figure 12.4. In the first step (a) the starting material was equilibrated with the starting buffer B1 (0.025 M tris·HCl, 0.035 M NaCl, pH 8.8) and separated from low-molecular-weight compounds on a matrix of Trisacryl (highly crosslinked

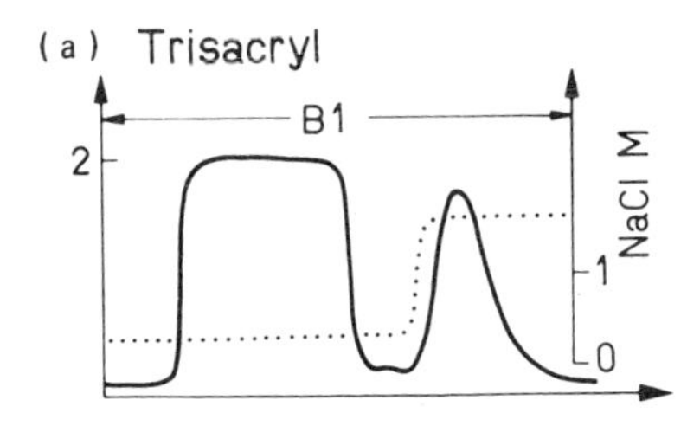

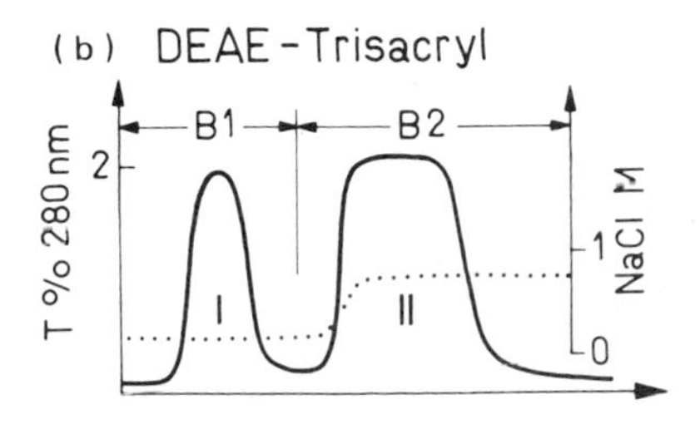

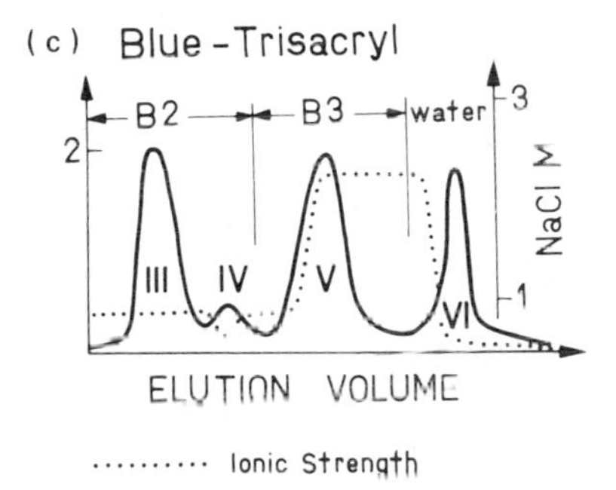

FIGURE 12.4 Elution profiles obtained during fractionation of human plasma cryosupernatant. (a) Desalting and equilibrium on Trisacryl. (b) Ion exchange on DEAE-Trisacryl. (c) Affinity chromatography on Blue-Trisacryl. (From Saint-Blancard et al., 1982.)

copolymer of N-[tris-(hydroxymethyl)methyl]acrylamide. In the second step (b) all plasma proteins were adsorbed to DEAE-Trisacryl M except for IgGs (I), which were eluted with high purity (99.5%). The adsorbed proteins were desorbed with buffer B2 (0.05 M tris·HCl, 0.75 M NaCl, pH 8) peak II. This peak II was then fractionated with

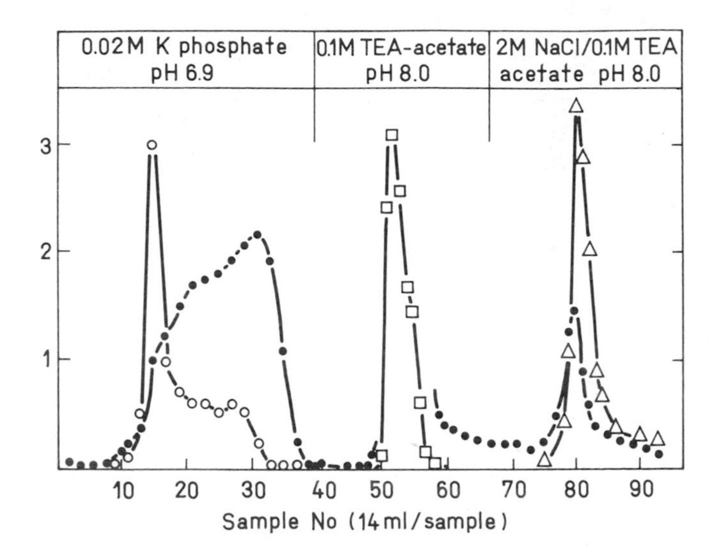

FIGURE 12.5 Separation of hemoglobin, catalase, and lactate dehydrogenase from human erythrocytes (2.9 mg protein) by chromatography on Blue-Sepharose. One interval along the vertical axis corresponds to $1A_{280}$ unit (●), 5000 units of catalase per milliliter (○), 20 mg HbO_2 per milliliter (□), and 0.5 units of lactate dehydrogenase per milliliter, determined fluorimetrically as NADH formation (△). (From Porumb et al., 1982.)

Blue Trisacryl (c). Under these conditions albumin and lipoproteins were adsorbed. The other proteins did not elute uniformly. The first peak (III) mainly contains transferrin and the second peak (IV) α- and β-globulins in low quantities. Albumin (V) was desorbed with a purity superior to 99% with buffer B3 (0.05 M tris·HCl, 3.5 M NaCl, pH 8). The yield of albumin is higher than 80%. The lipoproteins can be eluted with distilled water (VI). An interesting application was found by Porumb and coworkers (1982). The hemolysate from washed erythrocytes contains, beyond oxyhemoglobin, the catalase and glycolytic enzymes. For the preparation of highly purified hemoglobin the hemolysate was fractionated on Blue-Sepharose (Fig. 12.5). At pH 6.9 hemoglobin and cofactor-dependent enzymes were bound to the matrix, and the catalase and carbonic anhydrase – both with isoelectric points of 5.8 and 5.3 – were eluted with the starting buffer. Pure oxyhemoglobin was then eluted with buffer, pH 8, of low ionic strength. The other accompanying enzymes could be then eluted with buffer, pH 8, containing 2 M NaCl. The authors interpreted this behavior of Blue-Sepharose as the properties of an ion-exchanger combined with its specific affinity to dinucleotide-requiring enzymes.

Interferons are a heterogenous group of biologically active macromolecules. Their impurities vary depending on cell type, stimulation, culture conditions, and other factors. Therefore, affinity chromatography appears to be a suitable method. The variety of methods devised has been described elsewhere (Table 5.3). Blue-Sepharose CL-6B or Dextran Blue-Sepharose have been used to purify interferon from a wide range of sources (Jankowski et al., 1976; Bollin et al., 1978; Erickson and Paucker, 1979; Knight and Fahey, 1981).

Human lymphoid interferons were purified more than 1000-fold by a modified acid ethanol extraction and chromatographic procedure on Blue-Sepharose, and fibroblast interferons were purified completely. The nature of the interaction is not clear in detail. Any results show that the anthraquinone group is generally essential. If the dye is bound to the matrix via the primary amino group of the anthraquinone group, its affinity to interferon decreases because this group is then blocked sterically.

Neame and Parikh (1982) have investigated the adsorption of human lymphoblastoid interferon to a variety of immobilized triazine dyes. More than 85% was adsorbed by Procion Red HE-7B, Red HE-3B, Violet H-3B, and Cibacron Blue F3G-A. They found that Sepharose-immobilized Procion Red HE-7B is the most suitable gel for preparative-scale purification. From cell supernatant, 10^5 reference units per milliliter of interferon are adsorbed and can be eluted with 25-fold purification and 90% yield with a KCl gradient.

OTHER DYES

Stimulated by the unexpected properties of Cibacron Blue F3G-A as affinity ligand, a systematic search for further appropriate textile dyes began, particularly of the Cibacron and Procion series. The Procion Red HE-3B (Fig. 12.6), for instance, is a suitable dye that is already commercially available in matrix-bound form. This dye is a group-specific ligand similar to Cibacron Blue F3G-A but with different selectivity. It has a higher affinity to $NADP^+$-dependent dehydrogenases than the Cibacron Blue (Watson et al., 1978). Conversely, Cibacron Blue F3G-A has a higher affinity to NAD^+-linked enzymes. The interaction with the enzymes and proteins depends on steric factors and ionic and/or hydrophobic forces.

In addition to these two dyes, Amicon Corporation (Lexington, Massachusetts) offers three other matrices: Matrex Gel Orange A, Matrex Gel Green A and Matrex Gel Blue B. Partial structures are

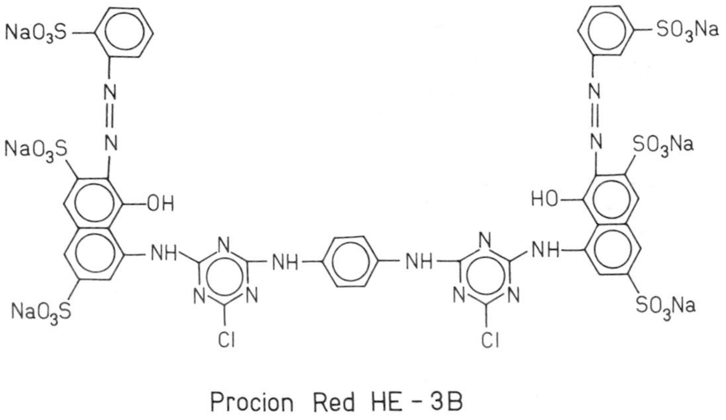

Procion Red HE - 3B

FIGURE 12.6 Chemical structure of Procion Red HE-3B.

given in Figure 12.7. The ligand of Matrex Gel Blue B is a copper phthalocyanine chromophore. The adsorption behavior of these ligands is different. Orange A interacts with few enzymes in relation to Cibacron Blue and Procion Red, but it has a high degree of specificity, for lactate dehydrogenase from rabbit muscle and rat liver, for example. Green A binds many proteins, in contrast to Orange A and Blue B is similar to Orange A in its low binding properties (Amicon, 1980).

Anthraquinone dyes without a triazine group can also be used successfully as ligands. Mislovičová and colleagues (1980) have described the purification of rat liver lactate dehydrogenase with Remazol Brilliant Blue R (Fig. 12.7) as ligand and macroporous cellulose as matrix.

To clear up the relation between structure and affinity, reactive dyes were used that differ in the structural group. Gemeiner and coworkers (1981) have investigated the interaction between rabbit muscle lactate dehydrogenase and the following four dyes: Cibacron Blue F3G-A, Procion Blue MX-R, Remazol Brilliant Blue R, and Ostazin Brilliant Red S5-B (Fig. 12.7). The results show a substantial importance of the terminal benzene sulfonane moiety in the structure of Cibacron Blue and the involvement of the triazine group of Ostazin Red into affinity formation toward lactate dehydrogenase. Since the applied dyes are cheap and their coupling to suitable matrices can be done by simple procedures, one may expect that dye-ligand chromatography will gain further importance, especially for large-scale purification of enzymes and other proteins.

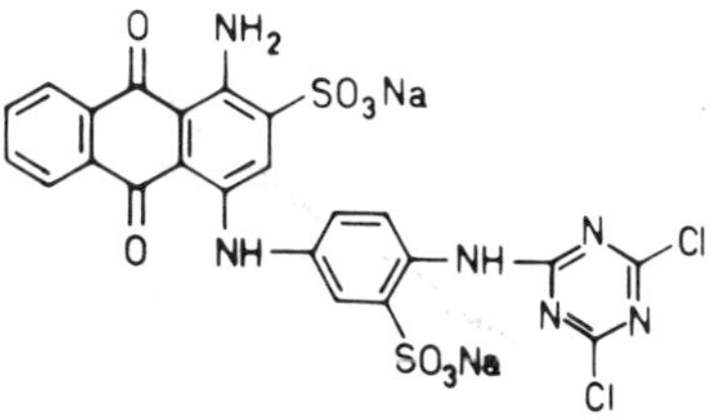

Procion Blue MX-3G

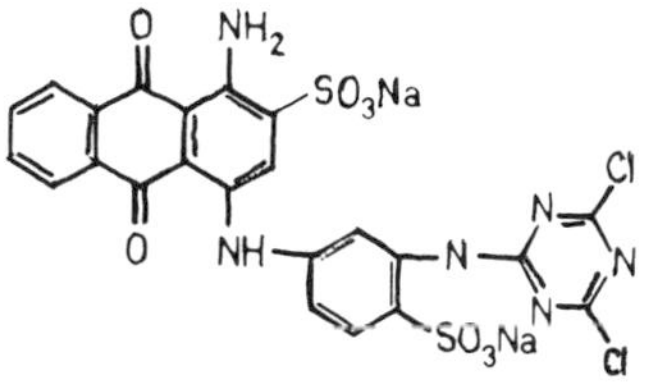

Procion Blue MX-R

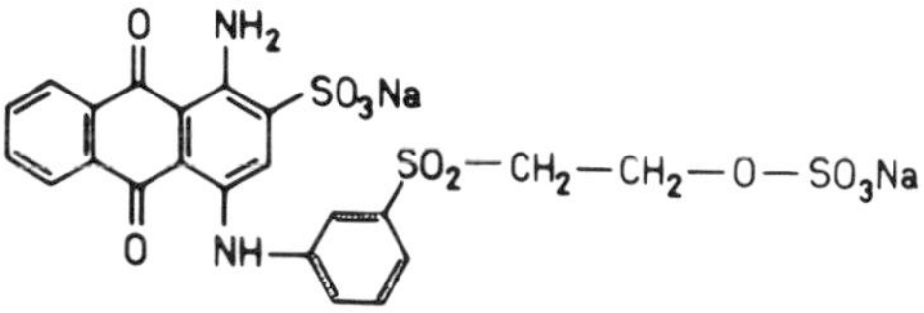

Remazol Brilliant Blue R

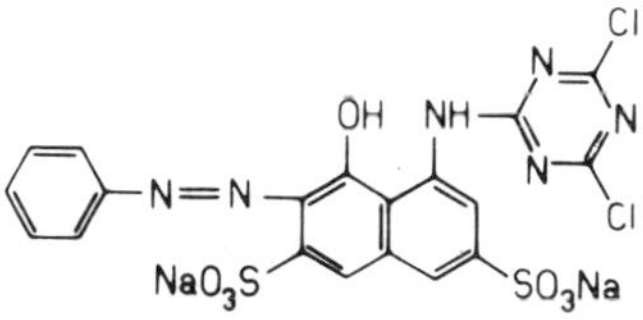

Ostazin Brilliant Red S-5B

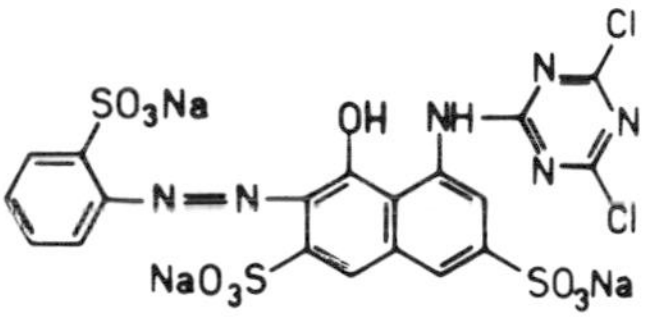

Procion Red MX-2B

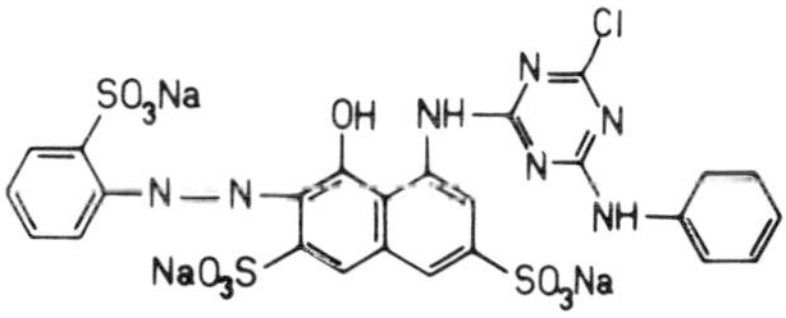

Procion Red H-3B

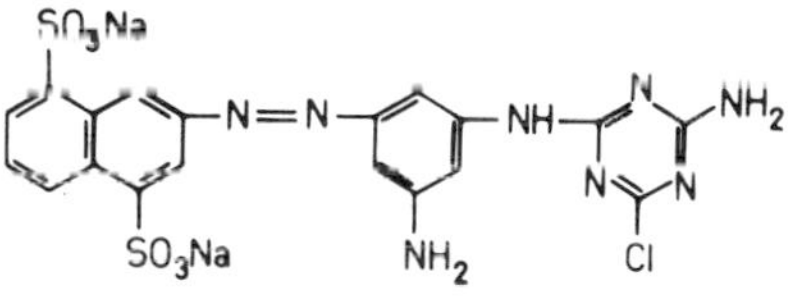

Orange A

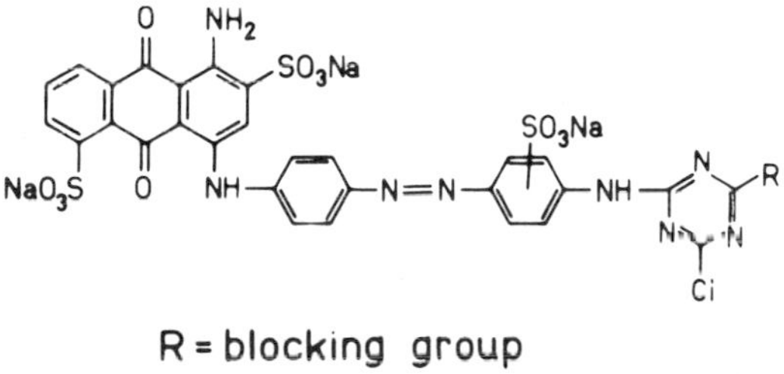

R = blocking group
Green A (possible partial structure)

FIGURE 12.7 Chemical structures of dye-ligands.

REFERENCES

Amicon Corporation (1980). *Dye-Ligand Chromatography*. Amicon Corporation, Lexington, Massachusetts.

Angal, S., and Dean, P. D. G. (1977). The effect of matrix on the binding of albumin to immobilized Cibacron Blue. *Biochem. J. 167*: 301-303.

Arnaud, P., and Gianazza, E. (1982). A two-step purification procedure for α_2-macroglobulin based on pseudoligand affinity chromatography. *FEBS Lett. 137*: 157-161.

Baksi, K., and Rushizky, G. W. (1978). Rapid purification of restriction endonucleases on Cibacron Blue F3G-A. *Fed. Proc., Fed. Am. Soc. Exp. Biol. 37*: 1414.

Baksi, K., and Rushizky, G. W. (1979a). Purification of the restriction endonuclease Pal I. *Anal. Biochem. 99*: 207-212.

Baksi, K., and Rushizky, G. W. (1979b). Purification of Pal I and its properties. *Fed. Proc., Fed. Am. Soc. Exp. Biol. 38*: 486.

Baksi, K., Rogerson, D. L., and Rushizky, G. W. (1978). Rapid, single-step purification of restriction endonucleases on Cibacron Blue F3G-A agarose. *Biochemistry 17*: 4136-4139.

Beissner, R. S., Quioche, F. A., and Rudolph, F. B. (1979). Dinucleotide fold proteins. Interaction of arabinose binding protein with Cibacron Blue F3G-A. *J. Mol. Biol. 134*: 847-850.

Birkenmeier, G. (1981). Untersuchungen über die Wechselwirkung von Serumproteinen und trägerfixierten Cibacronblau F3G-A und anderen Triazinfarbstoffen. Ph.D. Thesis, Section of Medicine, Karl-Marx-University, Leipzig.

Birkenmeier, G., Usbeck, E., Saro, L., and Kopperschläger, G. (1983). Triazine dye binding of human α-fetoprotein and albumin. *J. Chromatogr. 265*: 27-35.

Böhme, H.-J., Kopperschläger, G., Schulz, J., and Hofmann, E. (1972). Affinity chromatography of phosphofructokinase using Cibacron Blue F3G-A. *J. Chromatogr. 69*: 209-214.

Bollin, E., Vastola, K., Oleszek, D., and Sulkowski, E. (1978). The interaction of mammalian interferons with immobilized Cibacron Blue F3G-A: Modulation of binding strength. *Prep. Biochem. 8*: 259-274.

Bridges, M. A., Applegarth, D. A., Johannson, J., Davidson, A. G. F., and Wong, L. T. K. (1982). Isolation of pure active α_2-macroglobulin from small-scale plasma samples. *Clin. Chim. Acta 118*: 21-31.

Cottreau, D., Levin, M. J., and Kahn, A. (1979). Purification and partial characterization of different forms of phosphofructokinase in man. *Biochim. Biophys. Acta 568*: 183-194.

Dean, P. D. G., and Watson, D. H. (1979). Protein purification using immobilised triazine dyes. *J. Chromatogr. 105*: 301-319.

Erickson, J., and Paucker, K. (1979). Purification of acid ethanol extracted hyman lymphoid interferons by Blue Sepharose chromatography. *Anal. Biochem. 98*: 214-218.

Gee, A. P., Borsos, T., and Boyle, M. D. P. (1979). Interaction between components of the human classical complements pathway and immobilized Cibacron Blue F3G-A. *J. Immunol. Methods 30*: 119-126.

Gemeiner, P., Mislovičová, D., Zemek, J., and Kuniak, L. (1981). Antraquinone-triazine derivatives of polysaccharides. Relation between structure and affinity to lactate dehydrogenase. *Collect. Czech. Chem. Commun. 46*: 419-427.

Gianazza, E., and Arnaud, P. (1982). A general method for fractionation of plasma proteins. Dye-ligand affinity chromatography on immobilized Cibacron Blue F3G-A. *Biochem. J. 201*: 129-136.

Gordon, G. L., and Doelle, H. W. (1976). Purification properties and immunological relationship of L(+)-lactate dehydrogenase from lactobacillus casei. *Eur. J. Biochem. 67*: 543-555.

Haeckel, R., Hess, B., Lauterborn, W., and Wuster, K.-H. (1968). Purification and allosteric properties of yeast pyruvate kinase. *Hoppe-Seyler's Z. Physiol. Chem. 349*: 699-714.

Haff, L. A., and Easterday, R. L. (1978). Cibacron Blue-Sepharose: A tool for general ligand affinity chromatography. In *Theory and Practice in Affinity Chromatography*, F. Eckstein and Sundaram (eds.). Academic Press, New York, p. 23.

Hägele, E., Neeff, J., and Mecke, D. (1978). The malate dehydrogenase isoenzymes of saccharomyces cerevisiae. *Eur. J. Biochem. 83*: 67-76.

Harvey, M. J. (1980). The application of affinity chromatography and hydrophobic chromatography to the purification of serum albumin. In *Methods of Plasma Protein Fractionation*, J. M. Curling (ed.). Academic Press, London, pp. 189-200.

Jankowski, W. J., von Muenchhausen, W., Sulkowski, E., and Carter, W. A. (1976). Binding of human interferon to immobilized Cibacron Blue F3G-A. The nature of molecular interaction. *Biochemistry 15*: 5182-5187.

Kelleher, P. C., Smith, C. J., and Pannell, R. (1979). Chromatography of non-human albumins on Cibacron Blue agarose. Application to the separation of albumin from rat alpha-fetoprotein. *J. Chromatogr. 173*: 415-418.

Knight, E., Jr., and Fahey, D. (1981). Human fibroblast interferon. An improved purification. *J. Biol. Chem. 256*: 3609-3611.

Kopperschläger, G., Freyer, R., Diezel, W., and Hofmann, E. (1968). Some kinetic and molecular properties of yeast phosphofructokinase. *FEBS Lett. 1*: 137-141.

Kopperschläger, G., Diezel, W., Freyer, R., Liebe, S., and Hofmann, E. (1971). Wechselwirkungen der Hefe-Phosphofructokinase mit Dextranblau 2000. *Eur. J. Biochem. 22*: 40-45.

Kopperschläger, G., Böhme, H.-J., and Hofmann, E. (1982). Cibacron Blue F3G-A and related dyes as ligands in affinity chromatography. *Advan. Biochem. Eng.* *25*: 101-138.

Leatherbarrow, R. J., and Dean, P. D. G. (1980). Studies on the mechanism of binding of serum albumin to immobilized Cibacron Blue F3G-A. *Biochem. J.* *189*: 27-34.

Mahany, T., Khirabadi, B. S., Gersten, D. M., Kurian, P., Ledley, R. S., and Ramwell, P. W. (1981). Studies on the affinity chromatography of serum albumins from human and animal plasmas. *Comp. Biochem. Physiol.* *68B*: 319-323.

Matthes, D. (1979). Interpretation of nuclear magnetic resonance spectra for lactobacillus casei dihydrofolate reductase based on the X-ray structure of the enzyme-methotrexate-NADPH complex. *Biochemistry* *18*: 1602-1610.

Mislovičová, D., Gemeiner, P., Kuniak, L., and Zemek, J. (1980). Affinity chromatography of rat liver lactate dehydrogenase on the Remazol derivative of bead cellulose. *J. Chromatogr.* *194*: 95-99.

Naval, J., Calvo, M., Lampreave, F., and Pineiro, A. (1982). Interactions of different albumins and animal sera with insolubilized Cibacron Blue. Evaluation of apparent affinity constants. *Comp. Biochem. Physiol.* *71B*: 403-407.

Neame, P. J., and Parikh, J. (1982). Sepharose-immobilized triazine dyes as adsorbants for human lymphoblastoid interferon purification. *Appl. Biochem. Biotechnol.* *7*: 295-305.

Nikodem, V. M., Johnson, R. C., and Fresco, J. R. (1977). Interaction between blue dextran and aminoacyl-tRNA synthetase from bakers' yeast. *Fed. Proc., Fed. Am. Soc. Exp. Biol.* *36*: 822.

Pannell, R., Johnson, D., and Travis, J. (1974). Isolation and properties of human plasma α_1-proteinase inhibitor. *Biochemistry* *13*: 5439-5445.

Pommerening, K., Beneš, M., Štamberg, J., Ziegenbein, R., and Scheler, W. (1982). Preparation and properties of dye-ligand bead cellulose. 23rd Microsymposium: *Selective Polymeric Sorbents*, Prague.

Porumb, H., Lascu, I., Matinca, D., Oarga, M., Borza, V., Telia, M., Popescu, O., Jebeleanu, G., and Barzu, O. (1982). Separation of erythrocyte enzymes from hemoglobin by chromatography on Blue-Sepharose. *FEBS Lett.* *139*: 41-44.

Röschlau, P., and Hess, B. (1972). Affinity chromatography of yeast pyruvate kinase with Cibacronblau bound to Sephadex G-200. *Hoppe Seyler's Z. Physiol. Chem.* *353*: 441-443.

Saint-Blancard, J., Kirzin, J. M., Riberon, P., Petit, F., Foucart, J., Girot, P., and Boschetti, E. (1982). A simple and rapid procedure for large scale preparation of IgGs and albumin from human plasma by ion exchange and affinity chromatography. In *Affinity Chromatography and Related Techniques*, T. C. J. Gribnau, J.

Visser, and R. J. F. Nivard (eds.). Elsevier Scientific, Amsterdam, pp. 305-312.

Stellwagen, E. (1977). Use of Blue Dextran as a probe for the nicotinamide adenine dinucleotide domain in proteins. *Acc. Chem. Res. 10*: 92-98.

Subramanian, S., and Kaufman, B. T. (1980). Dihydrofolate reductases from chicken liver and lactobacillus casei bind Cibacron Blue F3G-A in different modes and at different sites. *J. Biol. Chem. 255*: 10587-10590.

Sugiura, M. (1980). Purification of the T4 DNA ligase by Blue Sepharose chromatography. *Anal. Biochem. 108*: 227-229.

Swart, A. C. W., and Henker, H. C. (1970). Separation of blood coagulation factors II, VII, IX and X by gel filtration in the presence of Dextran Blue. *Biochim. Biophys. Acta 222*: 692-695.

Thompson, S. T., Cass, K. H., and Stellwagen, E. (1975). Blue-Dextran-Sepharose: An affinity column for the dinucleotide fold in proteins. *Proc. Nat. Acad. Sci. U. S. 72*: 669-672.

Travis, J., and Pannell, R. (1973). Selective removal of albumin from plasma by affinity chromatography. *Clin. Chim. Acta 49*: 49-52.

Travis, J., Bowen, J., Tewksbury, D., Johnson, D., and Pannell, R. (1976). Isolation of albumin from whole human plasma and fractionation of albumin depleted plasma. *Biochem. J. 157*: 301-306.

Travis, J., Garner, D., and Bowen, J. (1978). Human α_1-antichymotrypsin: Purification and properties. *Biochemistry 17*: 5647-5651.

Vician, L., and Tishkoff, G. H. (1976). Purification of human blood clotting factor X by Blue Dextran agarose affinity chromatography. *Biochim. Biophys. Acta 434*: 199-208.

Virca, G. D., Travis, J., Hall, P. K., and Roberts, R. C. (1978). Purification of human α_2-macroglobulin by chromatography on Cibacron Blue Sepharose. *Anal. Biochem. 89*: 274-278.

Watson, D. H., Harvey, M. J., and Dean, P. D. G. (1978). The selective retardation of NADP-dependent dehydrogenases by immobilized Procion Red HE-3B. *Biochem. J. 173*: 591-596.

Weber, B. H., Willeford, K., Moe, J. G., and Piszkiewicz, D. (1979). Hazards in the use of Cibacron Blue F3G-A in studies of proteins. *Biochem. Biophys. Res. Commun. 86*: 252-258.

Wille, L. E. (1976). A simple and rapid method for isolation of serum lipoproteins from the other serum protein constituents. *Clin. Chim. Acta 71*: 355-357.

13 Metal Chelate Affinity Chromatography

The basis of metal chelate affinity chromatography (Porath et al., 1975) is the ability of various metal ions (e.g., transition metals, Cu^{2+} and Zn^{2+}) to form coordination compounds with high- or low-molecular-weight ligands (Basolo and Johnson, 1964). The required adsorbents are prepared by introduction of chelate-forming groups (biscarboxymethylamino groups, salicylic acid, 8-hydroxyquinoline, amminosuccinic acid groups, and others) into a common gel followed by loading with a suitable metal ion, which has to be so strongly fixed that stable adsorption centers are formed. An essential prerequisite for the chromatographic effectivity of such polymeric coordination complexes in affinity chromatographic separation procedures is at least that (a) some of the metal-coordinated ligand molecules (e.g., H_2O, NH_3, and counterions) or ligand groups are exchangeable by metal complex-forming groups of the substance to be separated from the sample solution (see Fig. 2.1, Chap. 2), and (b) the metal ion has a much higher affinity for the gel than for this substance. The general run of a metal chelate chromatographic procedure using biscarboxymethylamino groups as the matrix-bound chelating ligand is illustrated schematically in Figure 13.1. To prevent emerging of metal ions in the effluent, it is advantageous to fill a part of the column (about one-third) with the metal-free gel (Lönnerdal et al., 1977).

Since chelate-forming ligands or groups normally are neutral or anion bases, the elution can be carried out by continuous or stepwise lowering of the pH. Nonspecific interactions may be suppressed by including a salt at high concentration in the buffer solution.

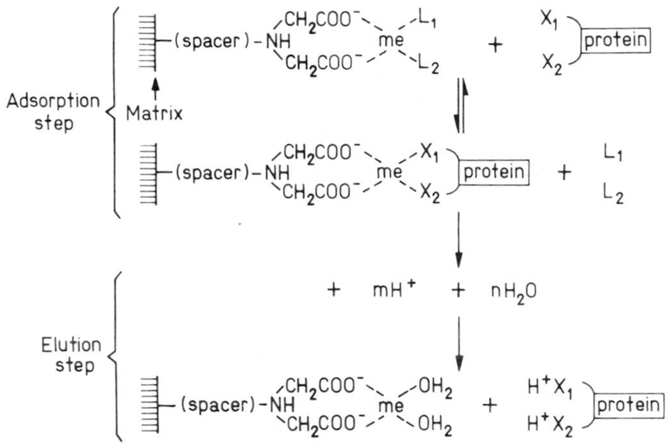

Adsorption step

Matrix —(spacer)–NH(CH_2COO^-)(CH_2COO^-) me (L_1)(L_2) + (X_1)(X_2) protein

⇌

Matrix —(spacer)–NH(CH_2COO^-)(CH_2COO^-) me (X_1)(X_2) protein + L_1 L_2

Elution step

↓ + mH^+ + nH_2O ↓

Matrix —(spacer)–NH(CH_2COO^-)(CH_2COO^-) me (OH_2)(OH_2) + (H^+X_1)(H^+X_2) protein

FIGURE 13.1 General run of metal chelate affinity chromatography.

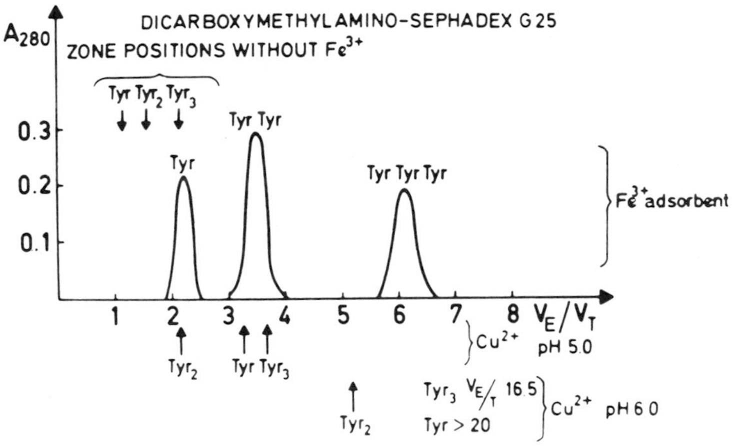

FIGURE 13.2 Schematic diagram indicating the position in terms of V_E/V_T units of tyrosine and di- and trityrosines on Fe^{3+} and Cu^{2+} (lower part) loaded with dicarboxymethylamino-Sephadex G25. All chromatograms were run on the same column with and without the metal ion indicated. The chromatographic experiments using the Fe^{3+} adsorbent and the metal-free gel were run in 100 mM ammonium acetate, pH 5.0. (From Porath, 1978.)

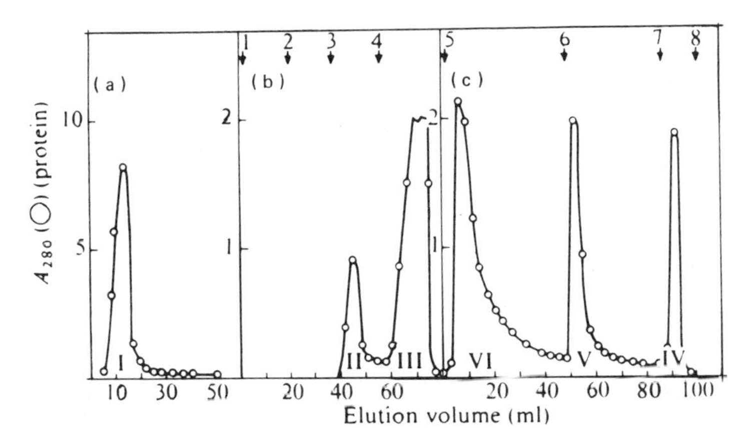

FIGURE 13.3 Composite chromatogram obtained by elution of the coupled Zn^{2+} and Cu^{2+} columns (a). The two columns were washed with 50 mM tris·HCl buffer, pH 8, 150 mM in NaCl (60 ml each). The adsorbed material was removed from each column separately according to the following scheme; Cu^{2+} column (b): (1) 100 mM Na phosphate buffer, pH 6.5, 18 ml; (2) 100 mM Na phosphate buffer, 800 mM in NaCl, pH 6.5, 20 ml; (3) 100 mM Na acetate buffer, 800 mM in NaCl, pH 4.5, 17 ml; (4) 50 mM EDTA, 50 mM in NaCl, pH 7.0, 25 ml. Zn^{2+} column (c): (5) 100 mM Na phosphate buffer, pH6.5, 47 ml; (6) 100 mM Na phosphate buffer, 800 mM in NaCl, pH 6.5, 38 ml; (7) 100 mM Na acetate, 800 mM in NaCl, pH 4.5, 12 ml; (8) 50 mM EDTA, 500 mM in NaCl, pH 7.0, 10 ml (see Table 13.1). (From Porath et al., 1975.)

Some examples demonstrating the efficiency of metal chelate chromatography are given in Figure 13.2. There it is shown that Fe^{3+} biscarboxymethylamino-Sephadex G25, contrary to the Fe-free gel, is very effective in the separation of Tyr, Tyr-Tyr, and Tyr-Tyr-Tyr (Porath, 1978). In Figure 13.2 is shown, moreover, the elution of these three compounds adsorbed on Cu^{2+} gel at pH 5 and 6.

The separation of various proteins from human serum was carried out using Zn^{2+} and Cu^{2+} biscarboxymethylamino-Sepharose CL 4B columns coupled in sequence (Porath et al., 1975). The result shown in Figure 13.3 and Table 13.1 demonstrates very impressively the separation power of metal chelate affinity chromatography for such biomacromolecules. In Table 13.2 are some selected examples showing that metal chelate affinity chromatography has already been used for successful isolation and purification of barely accessible peptides and proteins from materials of different origin.

TABLE 13.1 Components Identified by Gradipore Electrophoresis of the Fractions Obtained in Metal Chelate Affinity Chromatography of Human Serum on Cu^{2+} Column and Zn^{2+} Column

	Peak	Components identified
Material that passed both columns in conditions of adsorption:		
0.05 M tris·HCl, pH 8.0 15 M NaCl	I	Albumin
Material desorbed from Cu^{2+} column:		
Na acetate, 0.8 M NaCl, pH 4.5	II	Albumin, γ-globulins, prealbumin, traces of α_1-antitrypsin
0.05 N EDTA, 0.5 M NaCl, pH 7.0	III	Albumin, transferrin, haptoglobins, β-lipoprotein, traces of γ-globulins
Material desorbed from Zn^{2+} column:		
Na phosphate, pH 6.5	IV	Transferrin, α_1-antitrypsin, acid glycoprotein, γ-globulins, ceruloplasmin
Na phosphate, 0.8 M NaCl, pH 6.5	V	Transferrin, traces of hemoglobin, traces of γ-globulin
Na acetate, 0.8 M NaCl, pH 4.5	VI	α_2-Macroglobulin, traces of hemoglobin[a]

[a]Present as contaminant in the serum used.
Source: Porath et al. (1975).

Recently it was shown that metal chelate affinity chromatography offers further possibilities for protein separation when the chelate-forming groups and the metal ions, respectively, are varied (Porath and Ozin, 1983). In investigations using group IIIA metal ions (Al^{3+}, Ga^{3+}, In^{3+}, and Ti^{3+}) (Porath et al., 1983) it has been found that the separation characteristic of iminodiacetate- or tris(carboxymethyl)ethylenediamine-agarose columns for serum proteins depends widely on the nature of the chelate-forming metal ion. Further applications of metal chelate affinity chromatography can be expected in the near future.

TABLE 13.2 Examples of Substances Purified by Metal Chelate Affinity Chromatography

Substance	Adsorbent	Elution	References
Tyrosine-containing peptides	Fe^{3+} biscarboxymethylamino-Sephadex G25	100 mM ammonium acetate, pH 5.0	Porath (1978)
Human milk lactoferrin	Cu^{2+} biscarboxymethylamino-Sepharose 4B (50 μmol Cu/ml gel)	50 mM tris buffer/acetic acid + 500 mM NaCl, pH 2.8, or stepwise by the same buffer, pH 4.0	Lönnerdal et al. (1977)
Human plasma α_2-SH glycoprotein	Zn^{2+} biscarboxymethylamino-Sepharose 6B	Stepwise by different buffer systems containing NaCl	Lebreton (1977)
Human plasma α_2-macroglobulin	Zn^{2+} biscarboxymethylamino-Sepharose 4B and 6B	20 mM Na cacodylate buffer + 150 mM NaCl, pH 5.0	Kurecki et al. (1979)
α_1-Proteinase inhibitor		50 mM Na-phosphate buffer + 150 mM NaCl, pH 6.5	
Human fibroblast interferon	Zn^{2+} biscarboxymethylamino-Sepharose 6B	100 mM Na-acetate buffer + 1 M NaCl, pH gradient 6-4	Heine et al. (1981)
Squash NADH:nitrate reductase	Zn^{2+} biscarboxymethylamino-Sepharose 4B	75 mM K-phosphate (pH 6.2) + 1 M NaCl + 10 μM flavine adenine dinucleotide	Redinbaugh and Campbell (1983)

REFERENCES

Basolo, F., and Johnson, R. C. (1964). *Coordination Chemistry: The Chemistry of Metal Complexes*. W. A. Benjamin Inc., New York.

Heine, J. W., van Damme, J., Deley, M., Billian, A., and DeSomer, P. (1981). Purification of human fibroblast interferon by zinc chelate chromatography. *J. Gen. Virol. 54*: 47-56.

Kurecki, T., Kress, L. F., and Laskowski, Sr., M. (1979). Purification of human α_2 macroglobulin and α_1 proteinase inhibitor using zinc chelate chromatography. *Anal. Biochem. 99*: 415-420.

Lebreton, J. P. (1977). Purification of human plasma α_2-SH glycoprotein by zinc chelate affinity chromatography. *FEBS Lett. 80*: 351-354.

Lönnerdal, B., Carlsson, J., and Porath, J. (1977). Isolation of lactoferrin from human milk by metal chelate affinity chromatography. *FEBS Lett.* 75: 89-92.

Porath, J., and Olin, B. (1983). Immobilized metal ion affinity adsorption and immobilized metal ion affinity chromatography of biomaterials. Serum protein affinities for gel-immobilized iron and nickel ions. *Biochemistry 22*: 1621-1630.

Porath, J., Carlsson, J., Olsson, I., and Belfrage, G. (1975). Metal chelate affinity chromatography: A new approach to protein fractionation. *Nature 258*: 598-599.

Porath, J. (1978). Explorations into the field of charge transfer adsorption. *J. Chromatogr. 159*: 13-24.

Porath, J., Olin, B., and Granstrand, B. (1983). Immobilized-metal affinity chromatography of serum proteins on gel-immobilized group IIIA metal ions. *Arch. Biochem. Biophys. 225*: 543-547.

Redinbaugh, M. G., and Campbell, W. H. (1983). Purification of squash NADH nitrate reductase by zinc chelate affinity chromatography. *Plant Physiol. 71*: 205-207.

14 Charge Transfer Adsorption Chromatography

GENERAL CONSIDERATIONS

The basis of this chromatographic method is that certain pairs of organic compounds are able to interact under the formation of more or less stable molecular complexes that are essentially stabilized by an electron transfer from one component to the other. Besides the n-π transition, particularly π-donor acceptor interactions between π-electron donating and withdrawing aromatic or heterocyclic compounds, as demonstrated in Figure 14.1, are described in the literature. Numerous monographs and review articles dealing with the fundamentals of charge transfer interactions on the basis of the valence bond and molecular orbital method, as well as the free electron mode, have been published in the literature (Pfeiffer, 1937; Szent-György, 1960; Pullman and Pullman, 1960, Briegleb, 1961; Mulliken and Person, 1969; Slifkin, 1971). Aromatic and heterocyclic compounds with strong electron attracting groups, such as $-CH_2SO_2^-$, $-NO_2$, -CN, and =CO are effective π-electron acceptors, whereas electron releasing substituents, such as $-N(CH_3)_2$, $-OCH_3$, -OH, and CH_3 enhance the π-donor properties. A quantitative classification with regard to the electron donor acceptor properties is possible by means of the energy eigenvalues (k_j values) derived from an approximate solution of the Schrödinger equation:

$$E_j = \alpha + k_j\beta$$

where E_j is the electron energy in the j orbital, α the Coulomb integral, and β the exchange integral. Using the Hückel approximation,

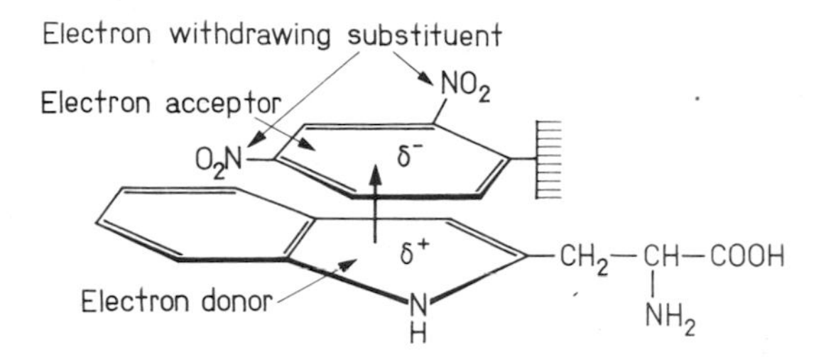

FIGURE 14.1 Schematic representation of charge transfer interaction.

electron energy diagrams for hundreds of biologically important compounds have been calculated by Pullman and Pullman (1960) and published in the literature. In Figure 14.2 some energy eigenvalues for compounds of interest in charge transfer adsorption chromatography are represented. A small k_j of the highest filled orbital indicates a good donor (e.g., tryptophan); a small k_j of the lowest vacant orbital characterizes a good acceptor (e.g., nitrophenyl derivatives). The stability of the charge transfer binding formed is determined finally by the energy difference between the highest occupied π-electron molecular orbital of the donor and the lowest empty molecular orbital of the acceptor molecule.

For the development of charge transfer adsorption chromatography as an applicable chromatographic method, fundamental investigations have been done by Porath and Dahlgren-Caldwell (1977). In separating any substance with electron donor or acceptor groups, a selected nonionizable compound with contrasting properties is chemically introduced in the matrix material. For this purpose sufficiently permeable uncharged and rigid hydrophilic supports, such as dextrans (Sephadex) or agaroses (Sepharose), are most suitable. Depending on whether the prepared adsorbent materials have electron releasing or withdrawing properties, electron acceptor gels and electron donor gels are distinguished in the literature (Porath, 1979).

Since a chromatographic procedure is carried out in aqueous solution, charge transfer interactions are never included exclusively in the donor acceptor complex formation, but they are overlapped more or less by hydrophobic attractions (and others, such as van der Waals-London forces, hydrogen bonds, electrostatic interactions, and molecular sieve effects (Porath and Dahlgren-Caldwell, 1977; Vijayalakshmi and Porath, 1979). The portion of such interactions in an affinity chromatographic procedure is determinant for the elution conditions. On the other hand, participation of real ionic interactions may be widely excluded by selection on nonionizable matrices and ligands.

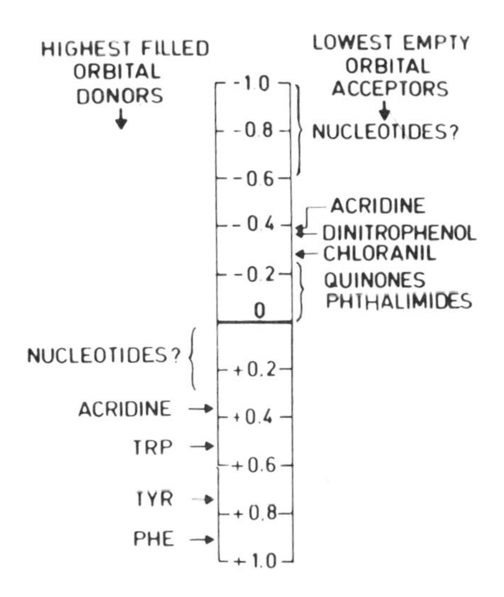

FIGURE 14.2 Charge transfer adsorption. Scale of energy eigenvalues (k_j values) from approximate solution of the Schrödinger equation (drawn from Pullman and Pullman, 1960). A small k_j of the highest filled orbital indicates a good donor; a small k_j of the lowest vacant orbital characterizes a good acceptor. (From Porath, 1978.)

In the following electron acceptor and electron donor gels are distinguished as proposed by Porath (1978), although practically an unambiguous assignment is often difficult by reason of the multiple forces and effects influencing the adsorption behavior already mentioned.

ELECTRON ACCEPTOR GELS

It is a prerequisite for the synthesis of acceptor gels that the acceptor ligand substances used be soluble in the reaction media and nonionizable and that the adsorbents are as much as possible negligibly hydrophobic. Considering these demands, Porath and Larsson (1978) have synthesized the gels represented in Figure 14.3 (see also Porath 1979). Among these, the type VI substance is the most stable, whereas the type I, III, and IV adsorbents are too labile and give a considerable leakage of the ligand even at pH 7. For experimental use, however, the type V adsorbent shows by far the most favorable behavior.

In Figure 14.4 is shown a typical elution diagram for the chromatographic resolution of an artificial mixture of dopamine, tyramine, serotonin, and indolacetic acid on dinitrophenyl Sephadex G25 at

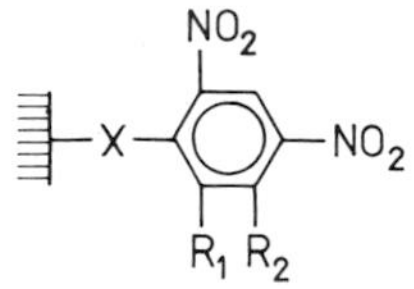

NO_2

X —— NO_2

R_1 R_2

	I	II	III	IV	V	VI
X	O	O	O	S	S	S-CH_2
R_1	NO_2	H	NO_2	NO_2	H	H
R	H	H	OH	H	H	H

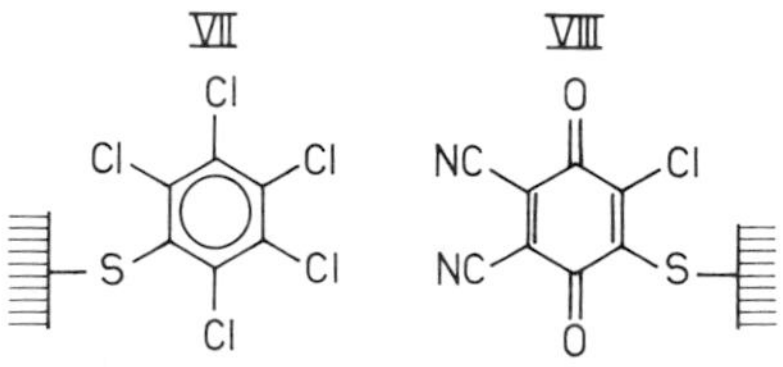

VII

Cl, Cl, Cl, Cl, Cl, S

VIII

O, NC, Cl, NC, S, O

FIGURE 14.3 Electron acceptor adsorbents.

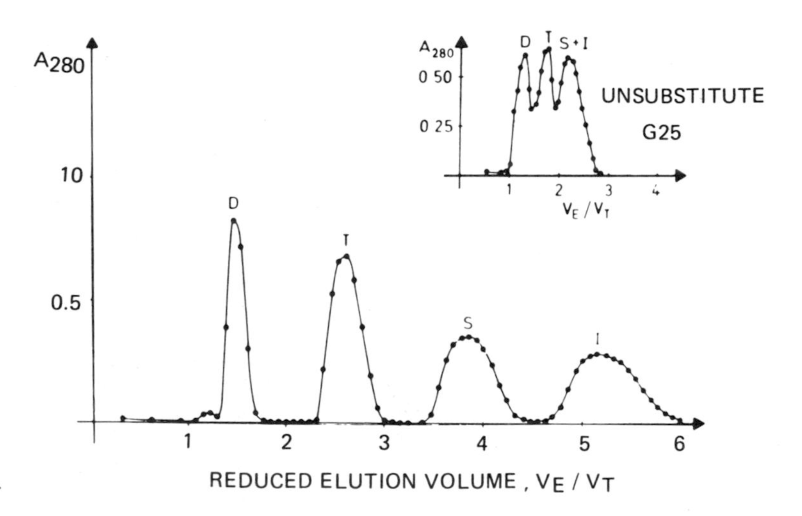

FIGURE 14.4 Chromatographic resolution of a mixture of dopamine (D), tyramine (T), serotonin (S), and indole acetic acid (I) on dinitrophenyl-S-Sephadex G25 at 4°C. Elution buffer: 100 mM ammonium formate, pH 3.2. Insert: pattern obtained on unsubstituted Sephadex G25 under identical operating conditions. (From Porath and Dahlgren-Caldwell, 1977.)

TABLE 14.1 Results of Experiments on DNP-S-Sepharose 6B (temperature, 25°C; pH 3.2)

Test substance	V_E/V_T	Test substance	V_E/V_T
Trp	1.29	Tyr-Tyr	1.41
Trp-Trp	5.27	Tyr-Tyr-Tyr	2.08
Trp-Gly	1.43	His-Tyr	1.29
Gly-Trp	1.51	His	1.16
Tryptamine-HCl	1.81	Phe	1.13
Indoleacetic acid	1.81	L-Dopa	1.16
Tyr	1.11	3-Iodotyrosine	1.34
Tyramine	1.25	3,5-Diiodotyrosine	1.94
Dopamine	1.31	3,5-Dibromotyrosine	1.25
Adrenaline	1.31	3,5-Dinitrotyrosine	1.11
Noradrenaline	1.29	Acetophenazine	13.75
Mescaline	1.28	Fluphenazine	9.82
Gly-Tyr	1.21	Chlorpromazine	15.25
Tyr-Gly-Gly	1.18		

Source: Porath and Dahlgren-Caldwell (1977).

4°C and pH 3.2, demonstrating the efficiency of this affinity chromatographic variant compared with the chromatographic behavior of un substituted Sephadex G25 as it is represented in the insert of Figure 14.4 (Porath and Dahlgren-Caldwell, 1977). The inclusion of 40% ethylene glycol in the elution medium decreases the adsorption only slightly (contrary to hydrophobic adsorption chromatography, in which an ethylene glycol gradient may be used to eluate adsorbed substances (see Chap. 15)), supporting the view that the solute-gel complex is stabilized by a combined effect of hydrophobic interactions and electron transfer adsorption. Further examples emphasizing the applicability of this method are given in Tables 14.1 and 14.2. Some data given in Table 14.1 support the assumption that charge transfer interactions essentially contribute to the complex stabilization (e.g., the strong increase of V_E/V_T from Trp to Trp-Trp). Finally,

TABLE 14.2 Results of Experiments on DNP-S-Sephadex G25

	V_E/V_T		
	DNP-S-Sephadex G25		Sephadex G25
Test substance	25°C	4°C	(25°C)
Phenylalanine	1.06		
Mescaline	1.10		
Dopamine	1.29	1.48	1.10
Tryptamine	1.99	2.60	1.45
Tryptophan	2.07		1.65
Serotonin	2.69	3.85	1.74
Indoleacetic acid	3.65	5.18	2.03

Source: Porath and Dahlgren-Caldwell (1977).

the degree of the adsorption of solutes on electron acceptor gels depends more or less strongly on the salt concentration, pH, and temperature. Moreover, Table 14.3 demonstrates with three different adsorbents (types V, VII, and VIII) that the adsorption power is strongly enhanced with increasing number of the electron donor centers in the substance to be separated (Porath, 1979). This behavior might indicate a risk for irreversible adsorption of proteins containing several superficially located donor substituents (e.g., tryptophan, tyrosine, and phenylalanine). At present, however, a theoretical interpretation of all these effects is still difficult.

Table 14.2 shows data obtained experimentally with dinitrophenyl Sephadex G25 (type V adsorbent) at 4 and 20°C. Control experiments were also carried out at 25°C with unsubstituted Sephadex G25. By contrast, an inclusion of 40% ethylene glycol into the elution medium decreases the adsorption only slightly. These results again confirm the assumption that electron charge transfer provides an essential contribution to the adsorption behavior of the substances mentioned.

ELECTRON DONOR GELS

It was found by Egly and Porath (1978, 1979), Egly and colleagues (1979), Porath (1978, 1979), and Rochette-Egly and coworkers (1979),

TABLE 14.3 Cooperative Adsorption Effect in 100 mM Formate

Ligand	Cl, Cl, Cl, Cl, Cl, S– VII	O, NC, Cl, NC, S–, O VIII	O_2N, S–, NO_2 V
Ligand concentration µmol/g	~400	~70	~200
Tyr	1.1	1.4	1.1
Tyr-Tyr	2.0	4.6	3.2
Tyr-Tyr-Tyr	V_E/V_T 4.6	9.1	7.4
Trp	2.8	3.3	2.7
Trp-Trp	19.0	30.0	30.0

Source: Porath (1979).

as well as Egly and Bochetti (1982), that aromatic ligands, such as acriflavin, anchored on Gels, provide efficient adsorbents in affinity chromatography of various substances, such as nucleotides and oligonucleotides and single-stranded and double-stranded nucleic acids. This is demonstrated in Figures 14.5 and 14.6, respectively, using

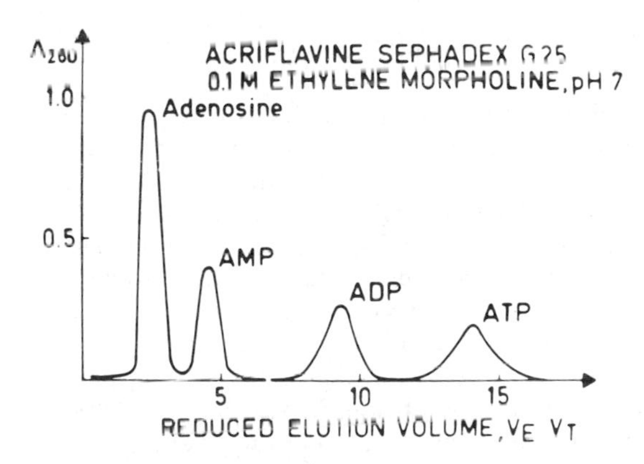

FIGURE 14.5 Fractionation of adenosine and adenine ribonucleoside phosphates (AMP, ADP, ATP: adenosine mono-, di-, triphosphate). (From Porath, 1978.)

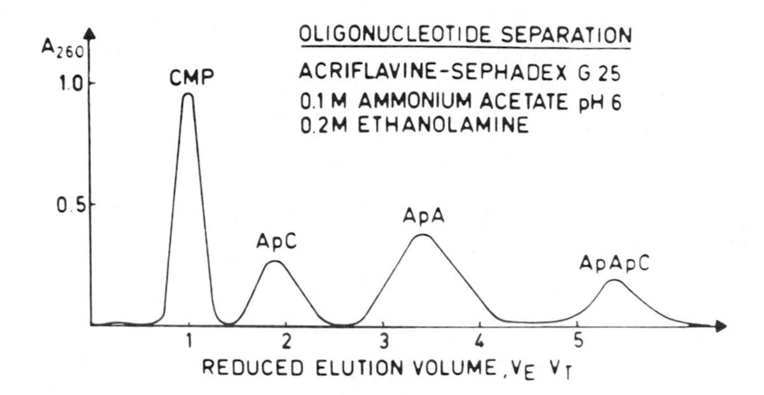

FIGURE 14.6 Chromatographic resolution of oligonucleotides: cytosine monophosphate (CMP), adenylylcytidine (ApC), adenylyladenosine (ApA), and adenylyladenylylcytidine (ApApC) on acriflavin-Sephadex G25 under different conditions. (From Porath, 1978.)

adenine ribonucleoside phosphate and oligonucleotide separations on adsorbent material of the following structure:

$$-OCH_2-\underset{OH}{\underset{|}{CH}}-CH_2O-(CH_2)_4-OCH_2-\underset{OH}{\underset{|}{CH}}-CH_2NH-(\text{acridinium, } \overset{+}{N})-NH_2$$

It was found, furthermore (Egly and Bochetti, 1982), that these acriflavin-substituted gels permit in some cases differentation between single-stranded and double-stranded nucleic acids (see also Chap. 9). Acriflavin itself is able to act both as electron acceptor and also as electron donor (Figure 14.2). In nucleotide separation, it is assumed that the electron donor properties of the gel contribute considerably to the adsorption (Porath, 1978, 1979), although the situation is not so clear as in the case of the acceptor gels (Egly et al., 1979). In contrast to electron acceptor chromatography, however, the forces participating in the adsorption of the solutes on acriflavin gels are more complex, as shown by the example of adenine nucleotide separation, in which increasing introduction of phosphate groups leads to enhanced V_E/V_T values (Table 14.4). This behavior refers additional ionic interactions between the adsorbent and the adsorbed nucleotide, additionally supported by the significant ionic strength influence on

TABLE 14.4 Influence of the Ionic Strength on the Acriflavin Adenine Related Compounds Interaction[a]

	V_E/V_T	
Compound	Ethylmorpholine-acetic acid, 200 mM (pH 7.0)	Ethylmorpholine-acetic acid, 200 mM (pH 7.0), 300 mM NaCl
Adenosine	3.32	3.27
Adenine	4.33	4.38
Cyclic AMP	7.12	3.51
AMP	7.22	2.07
ADP	14.92	1.97
ATP	28.99	1.83

[a]Samples (50 μl adenosine and adenosine nucleotides at 1 mg/ml) were run separately and simultaneously on acriflavin-Sephadex G25; column dimensions, 6.5 X 1 cm, flow rate, 15 ml/hr, temperature 24°C. Chromatography was followed spectrophotometrically. Each value is the average ±0.05 of three experiments.
Source: Rochette-Egly et al. (1979).

the adsorption behavior. It is difficult, therefore, to estimate unambiguously the significance of the electron donor properties of the gel material by its adsorption ability.

REFERENCES

Briegleb, G. (1961). *Elektronen-Donator-Acceptor Komplexe*. Springer Verlag, Berlin.

Egly, J.-M., and Bochetti, E. (1982). Immobilized acriflavin for aromatic interaction chromatography: Separation of nucleotides and nucleic acids. In *Affinity Chromatography and Related Techniques*, T. C. J. Gribnau, J. Visser, and R. J. F. Nivard (eds.). Elsevier Scientific, Amsterdam, pp. 445-451.

Egly, J.-M., and Porath, J. (1978). Gel matrices, coupling methods and charge transfer chromatography. In *Affinity Chromatography*, O. Hoffmann-Ostenhof, M. Breitenbach, F. Koller, D. Kraft, and O. Scheiner (eds.). Pergamon Press, Oxford, pp. 5-22.

Egly, J.-M., and Porath, J. (1979). Charge transfer and water mediated chromatography. II. Adsorption of nucleotides and related compounds on acriflavin-Sephadex, *J. Chromatogr. 168*: 35-47.

Egly, J.-M., Porath, J., Ochoa, J. L., Rochette-Egly, C., and Kempf, J. (1979). Isolation and interaction of aromatic compounds on aromatic substituted gels. In *Affinity Chromatography and Molecular Interactions*, J.-M. Egly (ed.). Editions INSERM, Paris, pp. 79-90.

Mulliken, R. S., and Person, W. B. (1969). *Molecular Complexes: A Lecture and Reprint Volume*. Wiley Interscience, New York.

Pfeiffer, P. (1937). *Organische Molekülverbindungen*, second edition. F. Enke, Stuttgart.

Porath, J. (1979). Charge transfer chromatography. *Pure Appl. Chem. 51*: 1549-1559.

Porath, J. (1978). Explorations into the field of charge transfer adsorption. *J. Chromatogr. 159*: 13-24.

Porath, J., and Larsson, B. (1978). Charge transfer and water mediated chromatography. I. Electron acceptor ligands on cross-linked dextran. *J. Chromatogr. 155*: 47-68.

Porath, J., and Dahlgren-Caldwell, K. (1977). Charge trans-adsorption chromatography. *J. Chromatogr. 133*: 180-183.

Pullman, B., and Pullman, A. (1960). Some electronic aspects of biochemistry. *Rev. Mod. Phys. 32*: 428-436.

Rochette-Egly, C., Kempf, J., and Egly, J.-M. (1979). A new chromatographic method using immobilized acriflavin for measurement cyclic AMP in cells prelabeled with radioactive adenine. *J. Cyclic Nucleotide Res. 5*: 397-406.

Slifkin, M. A. (1971). *Charge Transfer Interactions of Biomolecules*. Academic Press, London.

Szent-Györgyi, A. (1960). *Introduction to Submolecular Biology*. Academic Press, New York.

Vijayalakshmi, M. A., and Porath, J. (1979). Charge transfer and water mediated adsorption. III. Adsorption on tryptophan substituted Sephadex and Sepharose. *J. Chromatogr. 177*: 201-208.

15
Hydrophobic Interaction Chromatography

GENERAL CONSIDERATIONS

Hydrophobic interaction chromatography relies on the tendency of apolar molecules or groups to associate in aqueous solution. It has been pointed out by Tanford (1973) that the hydrophobicity of groups or molecules is proportional to their hydrocarbon-water interface. Experimentally it has been demonstrated that the association of apolar groups is thermodynamically characterized by an unusually high entropy contribution ΔS to the free energy ΔG of the whole aggregation, whereas the enthalpy contribution ΔH is low or even negative. This peculiarity is attributed to mobilization of ordered liquid water molecules structurally organized as clusters around the apolar groups, as schematically illustrated in Figure 15.1.

Structure-forming agents, such as certain electrolytes, tend to stabilize the water structure and, thus, to promote the association of apolar groups (von Hippel and Schleich, 1969; Jencks, 1969; Dandliker and DeSaussure, 1971), whereas structure-breaking or chaotropic electrolytes have a reverse effect on the water structure (Hamaguchi and Geiduschek, 1962; Hanstein et al., 1971).

Apolar groups occur relative frequently on the surface of protein molecules, such as phenylalanine and tryptophan (Klotz, 1970; Brand and Gohlke, 1972). Many proteins, therefore, are able to adsorb strongly apolar substances in aqueous milieu in a more or less entropically controlled reaction (Wishnia, 1969), suggesting a predominant contribution of hydrophobic interaction forces.

It has been shown in the last few years that various hydrogels substituted by apolar side chains of different structure and length (e.g.,

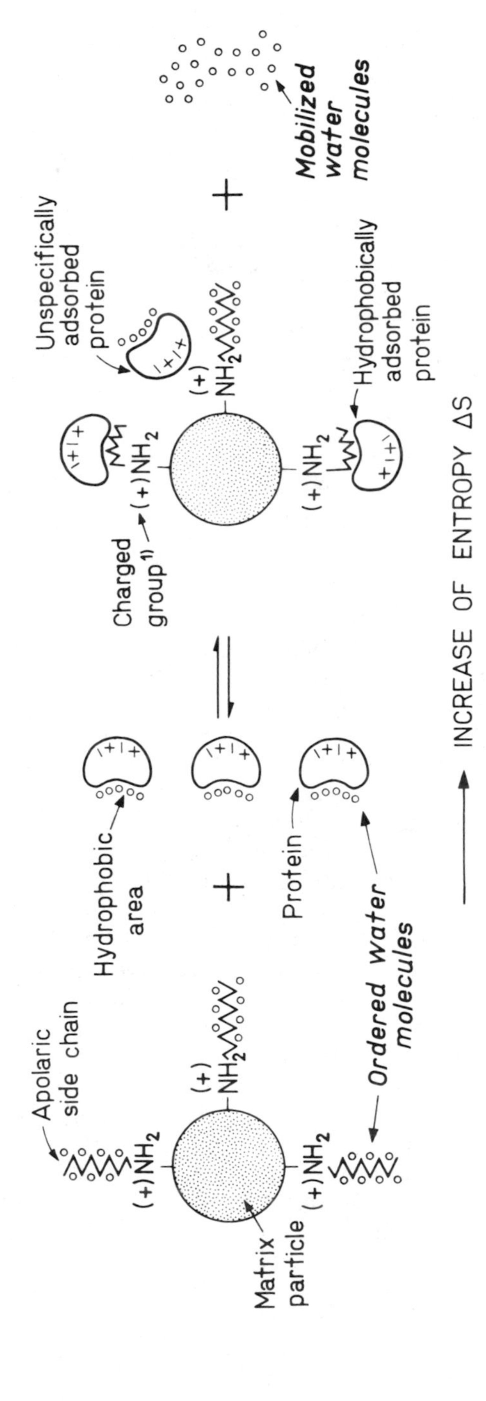

FIGURE 15.1 Schematic representation of hydrophobic interaction chromatography by alkaneamino adsorbents. pK = 9.7 from C_1 to C_{12}.

amino alkane groups) may be used successfully as ligands for liquid chromatographic separation of diverse proteins and nucleic acids in water as mobile phase (Yon, 1972; Er-el et al., 1972; Shaltiel and Er-el, 1973; Hofstee, 1973a, b; Hjerten, 1973; Hjerten et al., 1974; Jennissen and Heilmayer, 1975). On this basis hydrophobic interaction chromatography has been developed as a versatile routine method, and numerous review articles have been published in this field (Hjerten, 1976; Hofstee, 1976a, b; Hofstee and Otillio, 1978; Shaltiel et al., 1978; Jennisson, 1978; Palman, 1978; Hofstee, 1979).

Although, empirically, adsorption and elution can be adequately controlled, the molecular basis of interaction between the matrix material and the biomolecule to be separated has not been resolved completely. The theoretical understanding of processes connected with hydrophobic interaction chromatography essentially is complicated by the following. (a) the contact between partially apolar molecules or groups principally can occur by overlapping of a medley of interactions, such as van der Waals forces, charge transfer bonds, and hydrogen bridges but never hydrophobic interactions exclusively. (b) An adsorbed biomolecule can interact not only with the apolar ligands but also with other groups of the adsorbent. (c) As mentioned, the physical properties of water and consequently of the adsorbent-biomolecule contacts can be influenced considerably by various milieu factors, such as neutral salts, inorganic solutes, pH, temperature, and detergents. (d) Induced by the adsorption or by variation of the milieu, the conformation of biomacromolecules can be more or less changed and thereby the adsorption kinetics as well as the elution characteristics (Jennissen et al., 1982). For these reasons cooperative and overlapping effects must be considered generally in hydrophobic interaction chromatography. Various models have been proposed to interpret results and to have starting points for an optimization of adsorbent materials and chromatographic procedures. Shaltiel (1978) suggested that the adsorption in hydrophobic interaction chromatography is due to fitting the apolar alkyl residue of the matrix into an apolar pocket of the protein (Fig. 2.1, Chap. 2). In contrast, Jennissen and Heilmeyer (1975) demonstrated a correlation between apolarity of the gel and its adsorption properties. It could be shown by these authors that the binding of phosphorylase $\underline{b}$ and other proteins, such as phosphorylase kinase, 3',5'-cAMP-dependent protein kinase, and glycogen synthetase at low salt concentrations occurs as a function of the density of matrix substitution beginning at a density of alkyl residues, which is termed the "critical density" (Fig. 15.2). The capacity of the gels for phosphorylase kinase increases exponentially and reaches plateau values. By plotting the data in a double logarithmic form (see insert of Fig. 15.2), the sigmoidal curves can be linearized, in which their first part corresponds to the steep slope

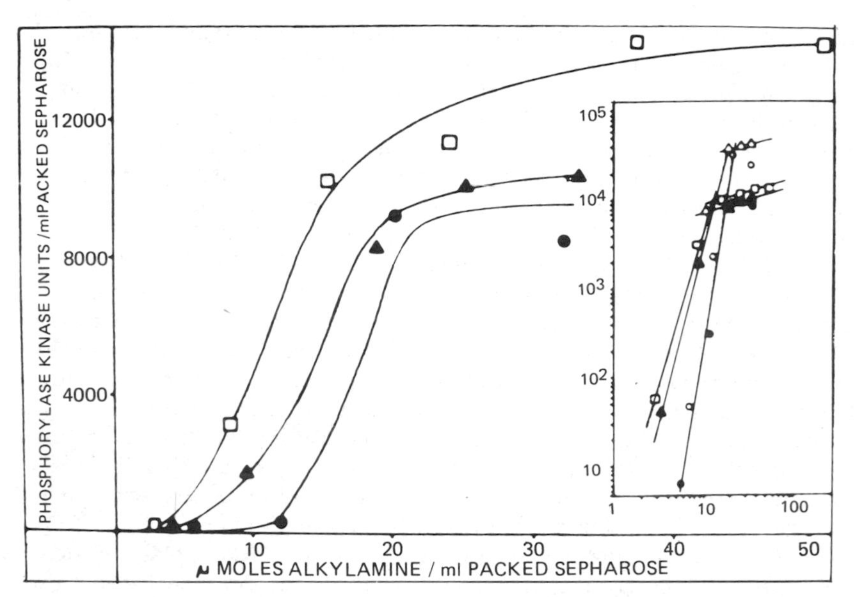

FIGURE 15.2 Adsorption of phosphorylase kinase on alkylamino Sepharose derivatives at 5°C as a function of the density of alkyl groups. The amount of adsorbed enzyme activity per milliliter of packed Sepharose was calculated from the difference between the total units applied and the units unadsorbed by the gel. Each column contained about 10 ml of packed Sepharose equilibrated with 10 mM sodium β-glycerophosphate, pH 7, 20 mM mercaptoethanol, 2 mM EDTA, 50 μM phenylmethylsulfonylfluoride, and 20% sucrose. Insert: double-logarithmic plots of adsorbed phosphorylase kinase as function of the degree of substitution. Phosphorylase kinase was applied in a crude extract (about 50 units/mg at pH 8.6). Unity of ordinates and abcissas is identical in both figures: methylamino-Sepharose (●) crude extract; (○) homogeneous enzyme. Ethylamine-Sepharose (▲) crude extract; (△) homogeneous enzyme. Butylamine-Sepharose (□) crude extract. (From Jennissen and Heilmeyer, Jr., 1975.)

whereas the plateau region yields low increments. From Hill plots it has been calculated that at least four to five alkyl residues of the matrix are involved in the adsorption of phosphorylase kinase and phosphorylase b, respectively. Therefore, it may be concluded that the adsorption of both enzymes does not follow a simple key-lock mechanism, as proposed for enzyme-substrate interactions, but is based on a positive cooperative interaction of the substituted alkyl residues, that is, on a multipoint interaction between the ligand and the substance to be separated (Jennissen, 1976a).

Elongation of the apolar substituents from one to four carbon atoms increases the hydrophobicity of the matrix and consequently the capacity for phosphorylase kinase and phosphorylase b rises about 60-fold (Er-el et al., 1972; Jennissen and Heilmeyer, 1975). The capacity of the matrix increases further by three to four magnitudes after enhancing the density of the apolar residues about threefold. That is, both the increase of the amino alkane group density at constant chain length and the elongation at constant density can provide the critical hydrophobicity needed to adsorb a protein.

EFFECTS OF NEUTRAL SALTS

Neutral salts may exert a relatively complex influence on the adsorption and elution of biomacromolecules in hydrophobic interaction chromatography (Pålman et al., 1977). The multiple and often different salt effects depending on type and concentration of the ions included may be understood on the basis of their water structure forming or breaking (chaotropic) properties, as well as on the basis of electrostatic interactions between the adsorbent material and the substance to be separated. On the basis of salt action, Jennissen (1976b) distinguishes two types of hydrophobic interaction chromatography. The first is "salting-in chromatography," which is characterized by the fact that a biomolecule is adsorbed at low salt concentrations and eluted at an increasing gradient chosen from the Hofmeister series. In Figure 15.3 the salting-in chromatography of phosphorylase kinase from ethylamine Sepharose with salt gradients of different ionic compositions is shown. The ionic strength of the peak fraction in Figure 15.3 is inversely related to the salting-in power of the ions employed. It may be assumed that this behavior is essentially attributed to the decreasing chaotropic potential of the applied ions from KSCN to Na_2SO_4. In the second type, "salting-out chromatography," the biomolecules are adsorbed at high ionic strength (>1 M) and eluted at decreasing salt concentration. In Figure 15.4 the chromatography of phosphorylase b using methylamine-Sepharose as adsorbent material is shown. From the result it can be supposed that the strong adsorption at a high salt concentration is essentially caused by the property of $(NH_4)_2SO_4$ to increase the water structure.

INFLUENCE OF ELECTROSTATIC CHARGES OF THE ADSORBENT MATERIAL

It has been stated, first by Shaltiel and Er-el (1973) and later by Hofstee (1976a), that the interaction of bovine serum albumin with

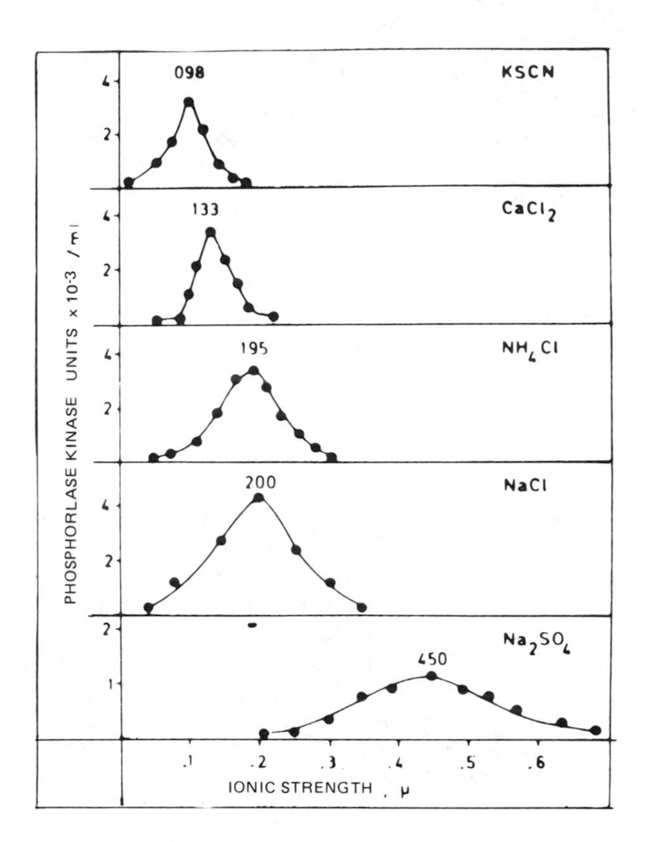

FIGURE 15.3 Desorption of purified phosphorylase kinase from ethylamine Sepharose (25 μm/ml) with salt gradients of different anionic composition. Each column with 5 ml of the gel was loaded with about 11 mg of purified phosphorylase kinase (see legend of Fig. 15.2). The linear gradients were produced from 100 ml of the adsorption buffer and 100 ml of salt-containing buffer. The ionic strength of these salts in the latter buffer was, for KSCN, 0.5; $CaCl_2$, 0.48; NH_4Cl, 0.75; NaCl, 1.5; and Na_2SO_4, 0.75. The gradients were controlled by measurements of the conductivity. The number at the maximum of the elution profiles indicates the ionic strength of the peak fraction. (From Jennissen and Heilmeyer, Jr., 1975.)

agaroses substituted by alkylamino groups of different chain length (sequence: agarose-NH-alkane) is reversed by the mere addition of NaCl, as demonstrated in Figure 15.5. There it is shown that the adsorption decreases at low salt concentrations (up to log [NaCl] = 0.5).

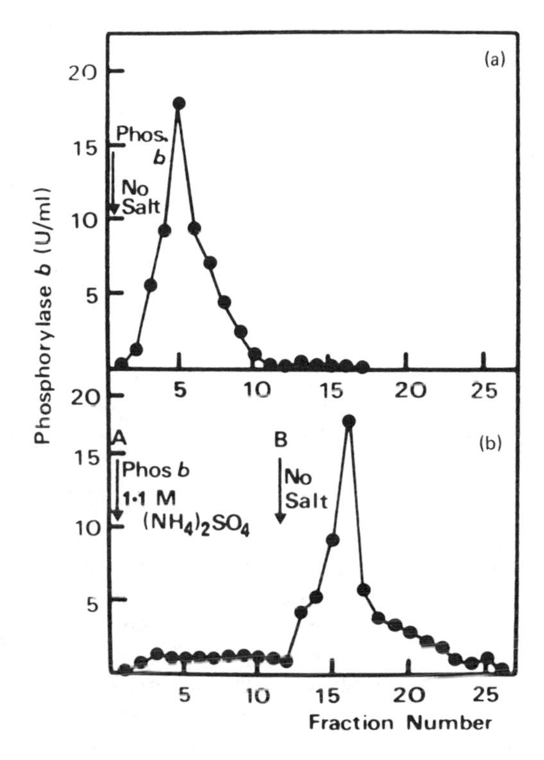

FIGURE 15.4 Salting-out chromatography of purified phosphorylase b (80 units/mg) on methylamine-Sepharose (30 μm/ml) at 5°C. Top panel: about 20 ml of the agarose was equilibrated with buffer (see legend to Fig. 15.2) in a column (15 X 2 cm). Phosphorylase b (6 mg per 3 ml) was applied (arrow). Fractions: 6.5 ml. Lower panel: about 20 ml of the agarose was equilibrated with buffer containing 1.1 ml ammonium sulfate in a column, as described. (a) Phosphorylase b (6 mg per 3 ml) in the equilibration buffer was applied. (b) Elution with ammonium sulfate-free buffer. (From Jennissen, 1976b.)

Since the -NH groups of the adsorbent materials are positively charged under the pH conditions given in Figure 15.5, it has been proposed that the adsorption takes place essentially via unspecific electrostatic interactions (Fig. 15.1). In this region of ionic strength, therefore, the binding capacity of the adsorbents utilized decreases with increasing salt concentration. A participation of more specific hydrophobic interactions may be assumed at NaCl concentrations >1 M, at which the binding strength is enhanced again, as shown in Figure 15.5, corresponding

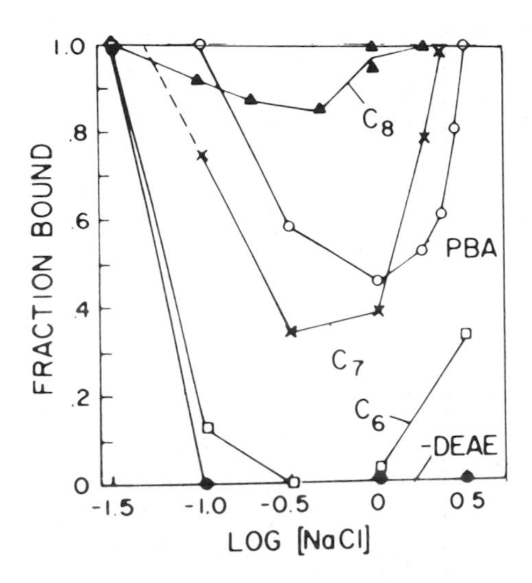

FIGURE 15.5 Effect of salt (NaCl) concentration on the fractional binding of bovine serum albumin by n-caprylamino (n-C_8)-, 4-phenyl-n-butylamino (PBA)-, n-heptylamino (n-C_7)-, n-hexylamino (n-C_6)-, or diethylaminoethyl (DEAE) agaroses. A few milligrams of the protein were applied to a 1 ml column of an adsorbent equilibrated with 0.01 M tris buffer·HCl (pH 8). The loaded column was washed exhaustively with the buffer alone and subsequently with buffer containing NaCl. (From Hofstee, 1976.)

to a salting-out effect. This interpretation was confirmed further by comparison of the protein binding capacity of different amino(n-alkane)-substituted adsorbents under saturation conditions. Figure 15.6a shows the data for the binding of bovine serum albumin, which is a relatively apolar protein. Moreover, the binding capacity of Ponceau S dye is included in this figure (Hofstee and Otillio, 1978). This dye can be used advantageously to determine the relative concentration of positive charges on the surface of matrix materials in the absence of neutral salts (Hofstee, 1976b). Figure 15.6b indicates that the hydrophobicity of the adsorbents utilized has little or no effect on the binding capacity for ovalbumin [against it, however, Pålman (1978) has shown that, with adsorbents of higher hydrophobicity, even ovalbumin is bound to a considerable extent]. Therefore, it must be assumed that electrostatic interactions play a dominant role in the adsorption of this protein. Contrary to this result the binding of bovine serum albumin is greatly affected above pentyl-substituted adsorbents,

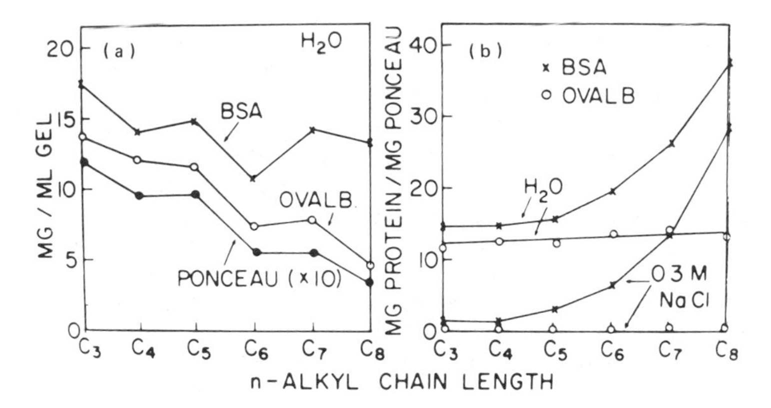

FIGURE 15.6 Effect of C chain length (C_3-C_8) of n-alkylamine-substituted agaroses on their binding capacities for bovine serum albumin (BSA) and ovalbumin (OV) as related to the capacities for binding of Ponceau S in absence of salt. After saturation of the adsorbents (about 1 ml) with protein in the presence of 0.01 M tris·HCl (pH 8) and exhaustive washing with water, the columns were first eluted with 10 ml of 0.3 M NaCl in 0.01 M tris·HCl, followed by elution with 10 ml of the same solution containing 50% DMF. (a) Binding capacities of the adsorbents for BSA, OV, and Ponceau S in the presence of water without additions. (b) Binding capacities in water and those in 0.3 M NaCl expressed as the amount of protein bound relative to the amount of Ponceau S bound in the absence of salt. (From Hofstee and Otillio, 1978.)

indicating that hydrophobic interactions are widely included. This interpretation is additionally supported by the effect of 0.3 M NaCl, which is attributed to a quenching of electrostatic interactions. For practical reasons it is always advantageous to select such milieu conditions in which a hydrophobic adsorption of a protein to be separated takes place, since hydrophobic interactions are much more selective as electrostatic contacts.

Moreover, the effectivity of a chromatographic procedure can be optimized in each case by chemical variation of the apolaric ligand (Fig. 15.6) and a convenient adaptation of the milieu.

PROTEIN SEPARATION ON HYDROPHOBICITY GRADIENTS

The elution of proteins bound to alkyl- or arylamino adsorbents with a gradient of decreasing ionic strength (e.g., diminishing the NaCl concentration) and polarity (e.g., by addition of ethylene glycol) seems not appropriate since adsorbents of arbitrarily chosen hydrophobicity may bind the more hydrophobic proteins of a mixture so fast that desorption is not possible without denaturation, whereas the less hydrophobic proteins may not be held by the adsorbents. For this reason, Hofstee and Otillio (1978) have proposed the use of hydrophobicity gradients. Such a gradient consists of a series of interconnected columns of increasing hydrophobicity through which the protein-containing solution is given, starting with the least hydrophobic. Each protein tends to be held by the adsorbent that provides the minimum degree of hydrophobicity required for binding. After washing, each disconnected column may be separately eluted by a selected mild eluent, such as a water-ethylene glycol mixture (Hofstee and Otillio, 1978).

As a practical example, the binding of serum proteins by a hydrophobicity gradient is represented in Figure 15.7. The data show that relatively little protein is adsorbed by the C_4 column. A peak appears at C_5, followed again by an increase in binding from C_6 to C_8. The sodium dodecylsulfate electrophoretic patterns of the eluates and of the protein solution after (rec) and before (Ser) recycling as well as of untreated solution of γ-globulin (Fr.II) is shown in the insert of Figure 15.7. The data indicate that the major component of the protein fraction bound by and eluted from the C_5-adsorbent consists of γ-globulin, whereas little γ-globulin is seen in the pattern of the C_8 eluate. It seems that the C_5 column is most suitable for the separation of the two major components, serum albumin and γ-glublin.

HYDROPHOBIC-IONIC CHROMATOGRAPHY

A further variant of hydrophobic interaction chromatography is the so-called hydrophobic-ionic chromatography described by Sasaki and colleagues (1982). In this method, Amberlite CG 50 possessing hydrophobic regions and ionizable carboxyl groups was used. At acidic pH ($\leq$4.5) all ion-exchange groups are protonated and the adsorption is widely due to hydrophobic interactions. At increasing pH the carboxyl groups become dissociated and the applied enzymes (glucose oxidase from *Aspargillus niger*, hyaluronidase from *Streptomyces hyalurolyticus*, and cholesterol oxidase as well as the esterase from *Pseudomonas fluorescens*) are eluted in a purified form with remarkable good yields.

CONCLUSIONS

This brief treatment of hydrophobic interaction chromatography shows that this method offers multiple possibilities for separation of biomolecules, including several nucleic acids. In most cases, however, it is

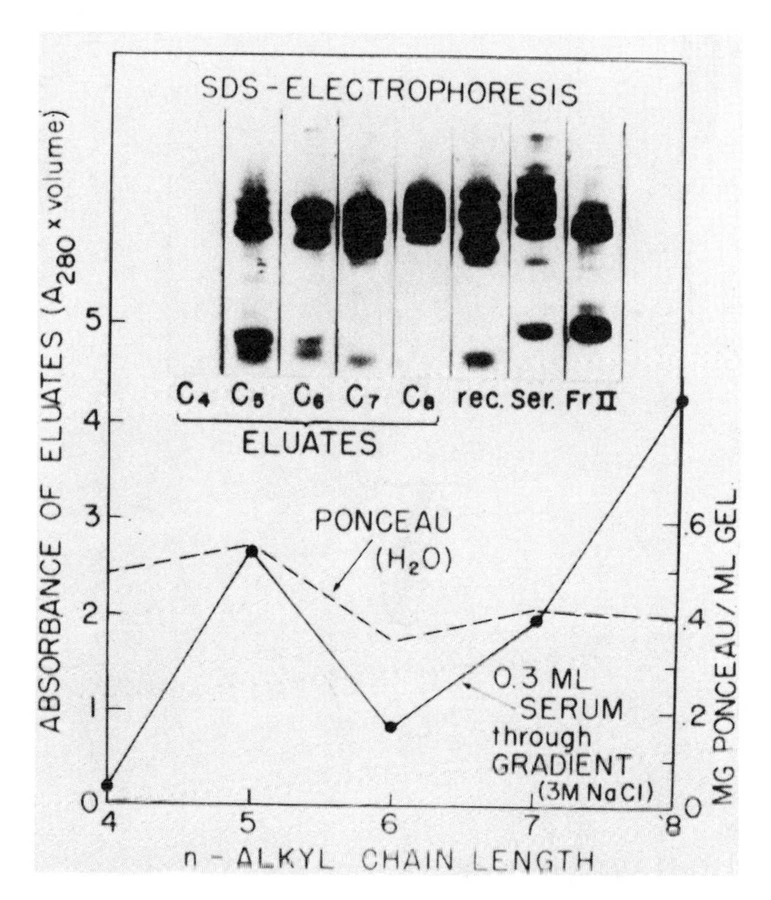

FIGURE 15.7 Adsorption chromatography of 0.3 ml human blood serum on a hydrophobicity gradient consisting of interconnected 1 ml columns of n-butyl (C_4)-, n-pentyl (C_5)-, n-hexyl (C_6)-, n-heptyl (C_7), and n-octyl (C_8)-amino agaroses prepared via CNBr activation from CL-Sepharose 4B. The adsorbents, preheated for 1 hr at 100°C and pH 4-5, were equilibrated at about 5°C with 3 M NaCl in 0.01 M tris·HCl (pH 8). The serum, diluted 1:50 with and dialyzed against the buffer salt medium, was recycled about 10 times through the gradient in the direction of increasing hydrophobicity. This was followed by exhaustive washing with the ambient medium and subsequent elution of the disconnected columns by means of 50% ethylene glycol in buffer containing 0.3 M NaCl. Before electrophoresis, the eluates of the C_4, C_5, C_6, C_7, and C_8 adsorbents were desalted and concentrated approximately 10-, 4-, 10-, 7-, and 2.5-fold, respectively. (From Hofstee and Otillio, 1978.)

TABLE 15.1 Examples of Substances Purified by Hydrophobic Interaction Chromatography

Substance	Matrix	Elution conditions	References
Maize leaf inorganic pyrophosphatase	Phenoxyacetyl cellulose	0.1% Triton X-100 or a t-butanol gradient	Dixon et al. (1979)
Chlorophylase	Phenyl-Sepharose CL-4B	Triton X-100 gradient (0-0.05%)	Shimokawa (1982)
Rabbit muscle glycogen debranching enzyme[a]	ω-Aminoethyl and butyl agarose	5 mM tris buffer + 1 mM EDTA + 50 mM NaCl, pH 7.0	White et al. (1981)
Prenyl transferases	C_0-C_{10} alkyl agaroses	Increasing Triton X-100 gradient (up to 0.5%)	Muth et al. (1979)
Porcin pancreatic enzymes (kallikrein, deoxyribonuclease, trypsin, carboxypeptidase A and B)	Amberlite CG 50	Increasing pH (4-6), different buffers	Sasaki et al. (1979)
Staphylococcal lipase	Octyl-Sepharose CL-4B	Triton X-100 gradient (0.1%), 50 mM tris·HCl buffer + 2 M KCl, pH 8	Jürgens and Huser (1981)

Heparin	Phenyl-Sepharose CL-4B	Ammonium sulfate gradient (3.8-2 M + 10 mM HCl)	Ogama et al. (1981)
Escherichia coli tRNA	Octyl-Sepharose CL-4B, naphthoyl-Sepharose CL-6B	Decreasing ionic strength beginning at 750 mM ammonium sulfate in 100 mM Na-phosphate buffer containing 10 mM Mg-acetate, pH 6.8 (when naphthoyl-Sepharose CL-6B is used as adsorbent)	Hjerten et al. (1979)
Nuclear androgen receptor	ω-Aminoalkyl agaroses (optimal chain length: (C_4, C_6)	Tes buffer[b] + 600 mM NaCl	Bruchovsky et al. (1981)
Porcine enteropathogenic *E. coli* strains	Phenyl and octyl Sepharose CL-4B (and others)	Decreasing ionic strength beginning at 4 M NaCl used in the adsorption step and elution with 10 mM Na-phosphate buffer, pH 6.8	Smyth et al. (1978)

[a]Containing the enzymes: 1.4-α-D-glucan: 1.4-α-D-glucan-4-α-glycosyltransferase (EC 2.4.1.25), and dextrin 6-α-D-glucosidase (EC 3.2.1.33).
[b]Tes buffer: 2-{[2-hydroxy-1,1-bis(hydroxymethyl)ethyl]amino}ethanesulfonic acid.

still necessary to ascertain the optimum structure of adsorbents empirically by testing. Nevertheless, in the meantime, the hydrophobic interaction chromatography has been developed to a manysided and powerful method in the separation of absolutely different biologically active substances, including microorganisms, as shown by the examples given in Table 15.1.

REFERENCES

Brand, L., and Gohlke, J. R. (1972). Fluorescence probes for structure. *Ann. Rev. Biochem. 41*: 843-868.

Bruchovsky, N., Rennie, P. S., and Comeau, T. (1981). Partial purification of nuclear androgen receptor by micrococcal nuclease digestion of chromatin and hydrophobic interaction chromatography. *Eur. J. Biochem. 120*: 399-405.

Dandliker, W. B., and DeSaussure, V. A. (1971). Stabilization of macromolecules by hydrophobic bonding: Role of water structure and chaotropic ions. In *The Chemistry of Biosurfaces*, Vol. 1, M. L. Hair (ed.). Marcel Dekker, New York, pp. 1-43.

Dixon, J., Andrews, P., and Butler, L. G. (1979). Hydrophobic esters of cellulose: Properties and applications in biochemical technology. *Biotechnol. Bioeng. 21*: 2113-2123.

Er-el, Z., Zaidenzaig, Y., and Shaltiel, S. (1972). Hydrocarbon-coated Sepharoses: Use in the purification of glycogen phosphorylase. *Biochem. Biophys. Res. Commun. 49*: 383-390.

Hamaguchi, K. M., and Geiduschek, E. P. (1962). The effect of electrolytes on the stability of the deoxyribonucleic helix. *J. Amer. Chem. Soc. 84*: 1329-1338.

Hanstein, W. G., Davis, K. A., and Hatefi, Y. (1971). Water structure and the chaotropic properties of haloacetates. *Arch. Biochem. Biophys. 147*: 534-544.

von Hippel, P. H., and Schleich, T. (1969). The effects of neutral salts on the structure and conformational stability of biological macromolecules in solution. In *Structure and Stability of Biological Macromolecules*, S. N. Timasheff and G. D. Fasman (eds.). Marcel Dekker, New York, pp. 417-474.

Hjerten, S. (1973). Some general aspects of hydrophobic interaction chromatography. *J. Chromatogr. 87*: 325-331.

Hjerten, S. (1976). Hydrophobic interaction chromatography of proteins on neutral adsorbents. In *Methods of Protein Separation*, Vol. 2, N. Catsimpoolas (ed.). Plenum, New York, pp. 233-243.

Hjerten, S., Hellman, U., Svensson, J., and Rosengren, J. (1979). Hydrophobic interaction chromatography of t-RNA. In *Affinity Chromatography and Molecular Interactions*, J.-M. Egly (ed.). Editions INSERM, Paris, pp. 315-320.

Hjerten, S., Rosengren, J., and Pålman, S. (1974). Hydrophobic interaction chromatography. The synthesis and use of some alkyl and aryl derivatives. *J. Chromatogr. 101*: 281-288.

Hofstee, B. H. J. (1973a). Hydrophobic affinity chromatography of proteins. *Anal. Biochem. 52*: 430-448.

Hofstee, B. H. J. (1973b). Protein binding by agarose carrying hydrophobic groups in conjugation with charges. *Biochem. Biophys. Res. Commun. 53*: 1137-1144.

Hofstee, B. H. J. (1976a). Hydrophobic effects in adsorptive protein immobilization. *J. Macromol. Sci., Chem. 10*: pp. 111-147.

Hofstee, B. H. J. (1976b). Hydrophobic adsorption chromatography of proteins. In *Methods of Protein Separation*, Vol. 2, N. Catsimpoolas (ed.). Plenum, New York, pp. 245-278.

Hofstee, B. H. J. (1979). Non-ionic effects in chromatographic separation of proteins through differential adsorptive immobilization. In *Affinity Chromatography and Molecular Interactions*, J.-M. Egly (ed.). Editions INSERM. Paris, 233-247.

Hofstee, B. H. J., and Otillio, F. (1978). Non-ionic adsorption chromatography of proteins. *J. Chromatogr. 159*: 57-89.

Jencks, W. P. (1969). *Catalysis and Enzymology*. McGraw-Hill, New York, pp. 417-436.

Jennissen, H. P. (1976a). Evidence for negative cooperativity in the adsorption of phosphorylase b on hydrophobic agaroses. *Biochemistry 15*: 5683-5692.

Jennissen, H. P. (1976b). Basic properties of hydrophobic agaroses. *Protides Biol. Fluids 23*: 675-697.

Jennissen, H. P. (1978). Cooperative phenomena in the adsorption of phosphorylase kinase and phosphorylase b on hydrophobic agaroses. In *Chromatography of Synthetic and Biological Polymers*, Vol. 2, R. Epton (ed.). Ellis Horwood Ltd., Chichester, pp. 45-52.

Jennissen, H. P., Demiroglou, A., and Logeman, E. (1982). Studies on the mechanism of protein adsorption on hydrophobic agaroses. In *Affinity Chromatography and Related Techniques*, T. C. J. Gribnau, J. Visser, and R. J. F. Nivard (eds.). Elsevier Scientific, Amsterdam, pp. 39-49.

Jennissen, H. P., and Heilmeyer, Jr., L. M. G. (1975). General aspects of hydrophobic chromatography. Adsorption and elution characteristics of some skeletal muscle enzymes. *Biochemistry 14*: 754-760.

Jürgens, D., and Huser, H. (1981). Large-scale purification of staphylococcal lipase by hydrophobic interaction chromatography. *J. Chromatogr. 216*: 295-301.

Klotz, J. M. (1970). Comparison of molecular structures of proteins: Helix content; distribution of apolar residues. *Arch. Biochem. Biophys. 138*: 704-706.

Muth, J. D., Baba, T., and Allen, C. M. (1979). Hydrophobic chromatography of prenyl transferases. *Biochim. Biophys. Acta 575*: 305-308.

Ogama, A., Matsuzaki, K., Uchiyama, H., and Nagasawa, K. (1981). Hydrophobic interaction chromatography of mucopolysaccharides: Examination of fundamental conditions for fractionation of heparin on hydrophobic gels. *J. Chromatogr. 213*: 439-451.

Pålman, S. (1978). Adsorption of proteins at high salt concentration on hydrophobically interacting matrices. In *Affinity Chromatography*, O. Hoffmann-Ostenhof, M. Breitenbach, F. Koller, D. Kraft, and O. Scheiner (eds.). Pergamon Press, Oxford, pp. 161-173.

Pålman, S., Rosengren, J., and Hjerten, S. (1977). Hydrophobic interaction chromatography on uncharged Sepharose derivatives: Effects of neutral salts on the adsorption of proteins. *J. Chromatogr. 131*: 99-108.

Sasaki, I., Gotoh, H., Yamamoto, R., Hasegawa, H., Yamashita, J., and Horio, T. (1979). Hydrophobic-ionic chromatography: Its application to purification of porcine pancreas enzymes. *Biochem. J. (Japan) 86*: 1537-1548.

Sasaki, I., Gotoh, H., Yamamoto, R., Tanaka, H., Takami, K., Yamashita, K., Yamashita, J., and Horio, T. (1982). Hydrophobic-ionic chromatography: Its application to microbial oxidase and cholesterol esterase. *Biochem. J. (Japan) 91*: 1555-1561.

Shaltiel, S. (1978). Hydrophobic chromatography. In *Chromatography of Synthetic and Biological Polymers*, Vol. 2, R. Epton (ed.). Ellis Horwood, Ltd., Chichester, pp. 13-41.

Shaltiel, S., and Er-el, Z. (1973). Hydrophobic chromatography: Use for purification of glycogen synthetase. *Proc. Nat. Acad. Sci. U. S. 70*: 778-781.

Shaltiel, S., Halperin, G., Er-el, Z., Tauber-Finkelstein, M., and Amsterdam, A. (1978). Homologous series of hydrocarbon-coated agarose in hydrophobic chromatography. In *Affinity Chromatography*, O. Hoffmann-Ostenhof, M. Breitenbach, F. Koller, D. Kraft, and O. Scheiner (eds.). Pergamon Press, Oxford, pp. 141-160.

Shimokawa, K. (1982). Hydrophobic chromatographic purification of ethylene enhanced cholorphyllase from citrus unshiu fruits. *Phytochem. 21*: 543-545.

Smyth, C. J., Jonsson, P., Olson, E., Söderlind, O., Rosengren, J., Hjerten, S., and Waldström, T. (1978). Differences in hydrophobic surface characteristics of porcine enteropathogenic Escherichia coli with or without K88 antigen as revealed by hydrophobic interaction chromatography. *Infect. Immunol. 22*, 462-472.

Tanford, C. (1973). *The Hydrophobic Effect: Formation of Micelles and Biological Membranes*. Wiley Interscience, New York.

White, R. C., Ruff, C. J., and Nelson, T. E. (1981). Purification of glycogen debranching enzyme from rabbit muscle using ω-aminoalkyl agarose chromatography. *Anal. Biochem. 115*: 388-390.
Wishnia, A. (1969). On the thermodynamic of induced fit. Specific alkane binding to proteins. *Biochemistry 8*: 5070-5075.
Yon, R. J. (1972). Chromatography of lyophilic proteins on adsorbents containing mixed hydrophobic and ionic groups. *Biochem. J. 126*: 765-767.

16

Covalent Chromatography (Chemisorption)

Covalent chromatography (Brocklehurst et al., 1973) is restricted to SH group-containing biologically active substances. The principle rests on the well-known fact that SH groups tend to slightly dimerize under formation of disulfide bridges and that this reaction is reversible under mild conditions.

In Figure 16.1 the general chemistry of covalent chromatography, including the regeneration step of the applied gels is shown. R-SH represents the free thiol group containing peptide or protein to be separated. In the first step (adsorption step), the substance to be separated is covalently bound to the chromatographic matrix. This involves selective attachment of thiol groups to the activated thiolated support by thiol-disulfide exchange with formation of a mixed disulfide. Since the coupling reaction is reversible after washing away unbound substances, an elution can be done by reduction of the disulfide bonds using low-molecular-weight thiol compounds, such as L-cysteine, mercaptoethanol, glutathione, and dithiothreitol.

When the thiol groups of the protein are unavailable for reaction due to steric shielding, as is the case in some proteins (e.g., parvalbumin from Hake muscle), the immobilization procedure must be performed in denaturing milieu (e.g., 8 M urea or 6 M guanidine hydrochloride) (Egorov et al., 1975). To isolate cysteinyl peptides containing their cysteine residues in disulfide form, the disulfides must first be reduced, as was described by Crestfield and colleagues (1963) for bovine pancreatic ribonuclease.

Mainly two types of matrices (as they have been developed by Pharmacia Fine Chemicals, Uppsala, Sweden) are used successfully

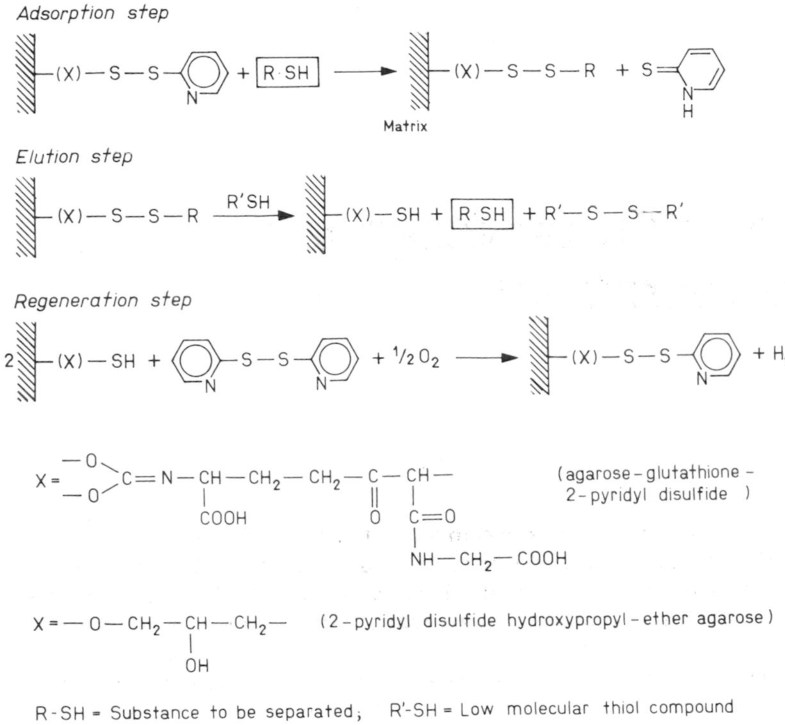

Adsorption step

$\text{Matrix}-(X)-S-S-\text{(2-pyridyl)} + \boxed{R\cdot SH} \longrightarrow \text{Matrix}-(X)-S-S-R + S{=}\text{(pyridine-2-thione, NH)}$

Elution step

$-(X)-S-S-R \xrightarrow{R'SH} -(X)-SH + \boxed{R\cdot SH} + R'-S-S-R'$

Regeneration step

$2\,-(X)-SH + \text{(2-pyridyl)}-S-S-\text{(2-pyridyl)} + {}^1\!/_2\,O_2 \longrightarrow -(X)-S-S-\text{(2-pyridyl)} + H_2O$

$X = (-O)_2C{=}N-CH(COOH)-CH_2-CH_2-C(=O)-CH(C{=}O\text{–}NH-CH_2-COOH)-$ (agarose-glutathione-2-pyridyl disulfide)

$X = -O-CH_2-CH(OH)-CH_2-$ (2-pyridyl disulfide hydroxypropyl-ether agarose)

R-SH = Substance to be separated; R'-SH = Low molecular thiol compound

FIGURE 16.1 General run of covalent chromatography.

in covalent chromatography: agarose glutathione-2-pyridine disulfide, in which the glutathione residue acts as spacer group (activated thiol Sepharose 4B), and 2-pyridyldisulfide-hydroxypropyl ether agarose with the hydroxypropyl group as hydrophilic spacer (thiopropyl Sepharose 6B). Both gels can be reactivated after the elution procedure by passing 1.5 mM solution of 2,2'-dipyridyldisulfide at pH 8, as shown in Figure 16.1.

The covalent chromatography has been used successfully for separation and purification of various thiol group-containing proteins and peptides, as well as of mercurated polynucleotides. Figure 16.2 is an elution diagram of the band 3 polypeptides of human erythrocyte membrane on a preparative scale. This procedure allows purification >95% (Kahlenberg and Walker, 1976). Table 16.1 shows a choice of further proteins purified by covalent chromatography, demonstrating the great efficiency of this variant.

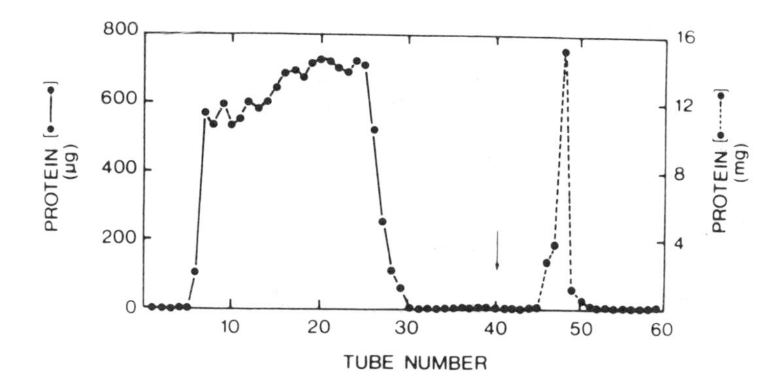

FIGURE 16.2 Preparative isolation of band 3 proteins from human erythrocyte membranes on activated thiol-Sepharose 4B. The Triton X-100 extract (60 ml) of membrane ghosts previously extracted with 2,3-dimethylmaleic anhydride was applied to a column (0.9 X 28 cm) of activated thiol-Sepharose 4B at a flow rate of 5.5 ml/hr and eluted with column buffer (300 mM NaCl, 1 mM EDTA, 0.5% Triton X-100 in 100 mM tris·chloride, pH 7.2). Arrow: Eluent supplement with 50 mM L-cysteine. (From Kahlenberg and Walker, 1976.)

Furthermore, the covalent chromatography offers two highly attractive possibilities for the isolation of thiol-containing peptides from proteins on a preparative scale (Carlsson et al., 1977). In the first method (Fig. 16.3a), the protein is immobilized through the free thiol groups followed by its proteolytic digestion. After washing away the unbound fragments, the peptide may be eluted with one of the reducing thiol reagents already described. In this manner the three cysteine-containing peptides from ceruloplasmin involved in the cupric ion binding have been isolated successfully (Egorov et al., 1975; Svenson et al., 1977).

The other way is to activate thiol groups on the protein by means of 2,2'-dipyridyldisulfide and then to digest the activated thiol protein in solution. The activated thiol peptides obtained may be purified by reversible immobilization on reduced covalent chromatography gels, as shown in Figure 16.3b.

Finally, covalent chromatography allows one to separate mercurated mRNA (Chap. 9). As has been shown by Dale and Ward (1975), cytosine and uracil residues in RNA and DNA can be mercurated without any loss of their functional behavior. Since mercury derivatives react with thiols to form mercaptides, sulfhydryl group-containing gels are suitable adsorbents for isolating both double and single

TABLE 16.1 Examples of Proteins Separated by Covalent Chromatography

Protein	Origin	Adsorbent material	References
Thiol-disulfide oxidoreductases	Beef liver	2-Pyridyldisulfide-hydroxypropyl ether agarose	Hillson and Freedman
Protein-disulfide isomerase			
Glutathion-insulin transhydrogenase			
Copper thioneine	Human fetal liver		Rydén and Duetsch (1978)
Band 3 proteins[a]	Human red cell membranes	Agarose glutathione-2-pyridyl-disulfide	Kahlenberg and Walker (1976)
γ-Globulin phospholipase[b]	Human serum		Scherer et al. (1976)
α-Chains of hemoglobin	Bovine hemoglobin		DeBruin et al. (1977)

[a]Purity >95%.
[b]Purification 10,000-fold.

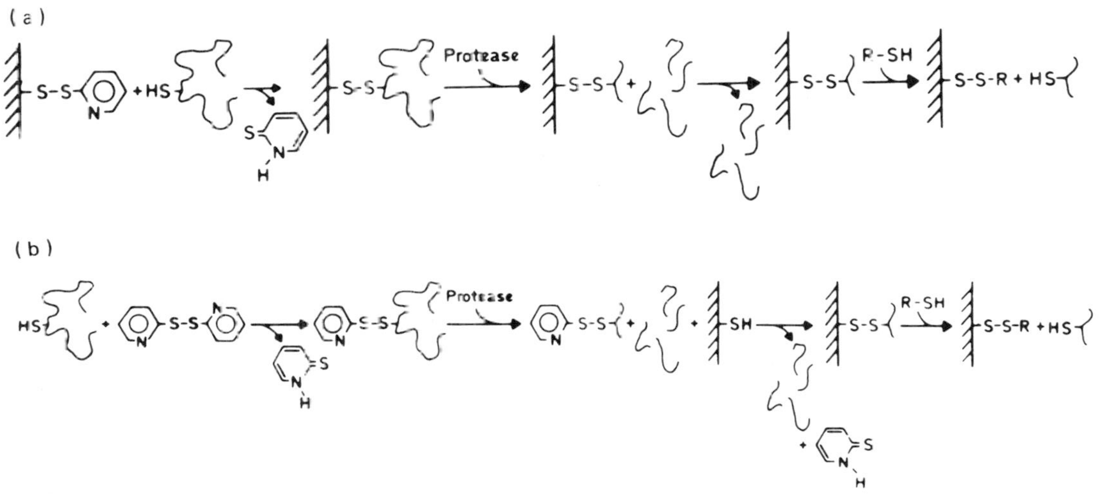

FIGURE 16.3 Isolation of thiol-containing peptides from protein by covalent chromatography. (a) Steps: (1) Immobilization of the thiol protein to solid-phase material (pH 2-9). (2) Digestion of the immobilized protein with a protease. (3) Washing out of the nonattached fragments. (4) Detachment of the thiol-containing peptides from the solid phase by reduction. (b) Steps: (1) Proteolytic digestion of the activated thiol protein. (2) Immobilization of the activated cysteinyl peptides to the solid-phase material. (3) Washing out of the nonattached fragments. (4) Detachment of the cysteinyl peptides from the solid phase by reduction (pH 8 with mercaptoethanol). (From Kahlenberg and Walker, 1976.)

polynucleotide strands. In their isolation of a specific gene structure, Georgiev and coworkers (1977) hybridized partially single-stranded DNA obtained from native DNA by exonuclease treatment with complementary mercurated mRNA. The entire DNA-mRNA complex was then purified on sulfhydryl agarose.

The efficiency of a covalent chromatographic procedure may be impaired by various side reactions, such as the leakage of gel-bound or protein-linked SH groups. In each of these cases the optimal conditions for the chromatographic procedure must be discovered (Carlsson et al., 1977).

REFERENCES

Brocklehurst, K., Carlsson, J., Kierstan, M. P. J., and Crook, E. M. (1973). Covalent chromatography. Preparation of fully active papain from dried papaya latex. *Biochem. J. 133*: 573-584.

Carlsson, J., Svenson, A., and Rydén, L. (1977). Isolation of cystein peptides from proteins by solid phase techniques based on thiol-disulfide exchange. In *Solid Phase Methods in Protein Sequence Analysis*. INSERM Symp. No. 5, A. Priviero, and M.-A. Coletti-Previero (eds.). Elsevier/North Holland Biomedical Press, Amsterdam, pp. 29-37.

Crestfield, A. M., Moore, S., and Stein, W. H. (1963). The preparation and enzymatic hydrolysis of reduced and S-carboxymethylated proteins. *J. Biol. Chem. 238*: 622-627.

Dale, R. M. K., and Ward, D. C. (1975). Mercurated polynucleotides: New probes for hybridization and selective polymer fractionation. *Biochemistry 14*: 2458-2469.

DeBruin, S. H., Joordens, J. J., and Rollema, H. S. (1977). An isolation procedure for the native α-chain of bovine hemoglobin. A study of the functional properties of this chain and its hybrid with the human β-chain. *Eur. J. Biochem.* 75: 211-215.

Egorov, T. A., Svenson, A., Rydén, L., and Carlsson, J. (1975). A rapid and specific method for isolation of thiol-containing peptides from large proteins by thiol-disulfide exchange on solid support. *Proc. Nat. Acad. Sci. U. S. 72*: 3029-3033.

Georgiev, G. P., Ilyin, Y. V., Ryskov, A. P., Tchurikov, N. A., Yenikolopov, G. N., Gvodez, V. A., and Ananiev, E. V. (1977). Isolation of eukaryotic DNA fragments containing structural genes and the adjacent sequences. *Science 195*: 394-397.

Hillson, D. A., and Freedman, R. B. (1979). Resolution of ox liver thiol:disulfide oxidoreductases by a new application of covalent chromatography. *Biochem. Soc. Trans.* 7, Part 3: 573-574.

Kahlenberg, A., and Walker, C. (1976). Preparative isolation of band 3, the predominant polypeptides of the human erythrocyte membrane. *Anal. Biochem. 74*: 337-342.

Rydén, L., and Duetsch, H. F. (1978). Preparation and properties of major copper-binding component in human fetal liver. *J. Biol. Chem. 253*: 519-524.

Scherer, R., Huber-Friedberg, W., Salem, A., and Ruhenstroth-Bauer, T. G. (1976). Phospholipase A_2 activity in human γ-globulin fraction. *Hoppe Seyler's Z. Physiol. Chem. 357*: 897-902.

Slykes, B. C. (1976). The separation of two soft-tissue collagens by covalent chromatography. *FEBS Lett. 61*: 180-185.

Svenson, A., Carlsson, J., and Eaker, D. (1977). Specific isolation of cysteine peptides by covalent chromatography on thiol agarose derivatives. *FEBS Lett. 73*: 171-174.

part V
RELATED TECHNIQUES

17
Affinity Partition

Certain biological macromolecules and cell particles have the property to spread in different concentrations in two nonmiscible aqueous phases (Albertsson, 1971). Such phases can be obtained by dissolving two different polymers, such as dextran and polyethylene glycol or polyethylene oxide in water (Shanbag and Johansson, 1974; Hubert et al., 1976). The partition of a given substance between two phases is determined by (a) the relative solvation of a substance in each of these phases, and (b) the electrical potential across the interface between them. The partition equilibrium, therefore, depends strongly on the kind and the concentration of the electrolytes included in the system (Shandhag and Johansson, 1974). Moreover, it is influenced by the pH and temperature, as well as by the concentration and molecular weights of the polymers and the length of their tie-line (Kroner et al., 1982). On the understanding that equal volumes of the two phases are compared, the ratio between the concentrations of biomacromolecules partitioned in the upper and lower phase is defined as the partition coefficient K.

The partition of pyruvate kinase and glutamate dehydrogenase is shown in Table 17.1 (Kopperschläger et al., 1983), indicating that these enzymes under the given conditions are enriched in the dextran-containing lower phase. When a ligand with affinity to the substance to be partitioned is introduced in one of the macromolecules (e.g., fatty acids and triazine dyes have been used successfully), the partition equilibrium is changed in the direction of that phase. This is illustrated in Figure 17.1 by pyruvate kinase and glutamate dehydrogenase, again where the polyethylene glycol is increasingly exchanged by a polyethylene glycol triazine conjugate (Kopperschläger et al.,

TABLE 17.1 Partition of Pyruvate Kinase and Glutamate Dehydrogenase in a Two-Phase System[a]

Enzyme	log K (n = 8)
Pyruvate kinase	-1.2 ± 0.10
Glutamate dehydrogenase	-0.48 ± 0.05

[a]Composed of 7% dextran T 500 (lower phase) and 5% polyethylene glycol (upper phase) expressed as the logarithm of the partition coefficient.

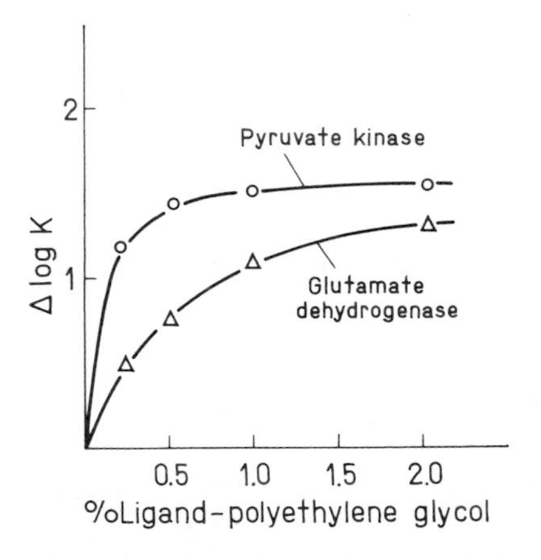

FIGURE 17.1 Alteration of the partition coefficient of pyruvate kinase and glutamate dehydrogenase depending on the concentrations of Procion Yellow HE-4R dye polyethylene glycol conjugate. The system contains 5% (w/w) polyethylene glycol 6000 with a different portion of dye-polyethylene glycol conjugate and 7% (w/w) Dextran T 500, 0.05 M sodium phosphate buffer (pH 7.0), 0.5 mM EDTA, and 5 mM 2-mercaptoethanol. After cooling to 0°C, about 1.5 units of each of the enzymes, dissolved in the same buffer, were added and the system was gently shaken for 15 sec. After separation of the phases samples of each of the two phases were removed for activity assay. The Δ log K was obtained from the difference in the log K values of the dye-conjugate-containing and the dye-conjugate-free system. (From Kopperschläger et al., 1983.)

1983). Since both enzymes have a considerable affinity to the dye-ligand, the enzymes are more and more enriched in the upper phase and attain a maximum value. Such a method, in which the partition of

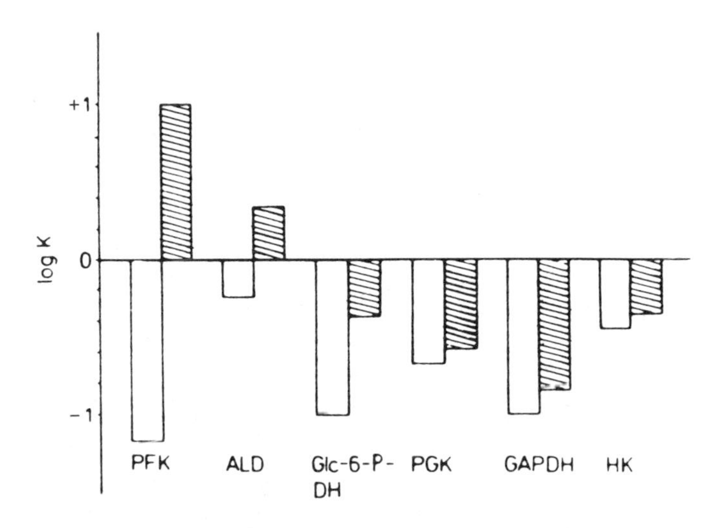

FIGURE 17.2 Partitioning of glycolytic enzymes in a biphasic system in the absence and presence of Cibacron Blue polyethylene glycol (Cb-PEG). A system containing 5% (w/w) PEG or Cb-PEG and 7.5% (w/w) dextran, 25 mM sodium phosphate buffer, pH 7.1, 0.5 mM EDTA, and 5 mM 2-mercaptoethanol was used. The systems were equilibrated at 0°C and then gently shaken for 15 sec. After separation of the phases at 0°C, the enzyme activities in both phases were determined. K: partition coefficient (ratio between activities in the upper and lower phase); PFK: phosphofructokinase; ALD: aldolase; Glc-6-P-DH: glucose-6-phosphate dehydrogenase; PGK: 3-phosphoglycerate kinase; GAPDH: glyceraldehydephosphate dehydrogenase; HK: hexokinase; White bars: without Cb-PEG; hatched bars: with Cb-PEG. (From Kopperschläger and Johansson, 1982.)

a substance between two nonmiscible liquid phases is controlled by its more or less specific adsorption to one of the solved polymer, is designated affinity partition (Flanagan and Barondes, 1975; Eriksson et al., 1976; Johansson et al., 1981). As the affine polymers, in most cases, triazine-dye polymer conjugates or fatty acid esters of polymers have been used up to now (Shandhag and Johansson, 1974; Johansson, 1976; Kula et al., 1979; Kopperschläger and Johansson, 1982).

In this chapter the general run of affinity partition is explained by the example of large-scale purification of phosphofructokinase from an extract of baker's yeast with polyethylene glycol-bound Cibacron Blue F3G-A (Kopperschläger and Johansson, 1982). Figure 17.2 shows the partition of a series of glycolytic enzymes in a biphasic polyethylene glycol and dextran-containing system in the presence and in the absence of Cibacron Blue polyethylene glycol conjugate. In the latter case all the enzymes are enriched in the lower phase. This phase is then separated and washed, and the enzyme phosphofructokinase is extracted by treatment with an upper phase containing 8.4% (w/w) Cibacron Blue polyethylene glycol conjugate instead of unsubstituted polyethylene glycol. After washing and addition of solid potassium phosphate for the separation of the enzyme, the product is concentrated by ion-exchange chromatography and purified by gel chromatography. The purification factor then amounts to about 140 compared with the cell homogenate.

Since various other enzymes are known to interact with triazine dyes (Dean and Watson, 1979; Kopperschläger et al., 1982) and these dyes are cheap and their covalent binding to polyethylene glycol is possible by simple nucleophilic reaction, it is assumed that affinity partitioning using dye-polymer conjugates will gain further importance in the future as a rapid and effective method for large-scale purification of a number of other proteins and enzymes (Lowe et al., 1982).

REFERENCES

Albertsson, P.-Å. (1971). *Partition of Cell Particles and Macromolecules*, 2nd ed. Almquist and Wikcell Stockholm; John Wiley & Sons, New York.

Dean, P. D. G., and Watson, D. H. (1979). Protein purification using immobilized triazine dyes. *J. Chromatogr. 165*: 301-319.

Eriksson, E., Albertsson, P.-Å., and Johansson, G. (1976). Hydrophobic surface properties of erythrocytes studied by affinity partition in aqueous two-phase systems. *Mol. Cell. Biochem. 10*: 123-128.

Flanagan, S. D., and Barondes, S. H. (1975). Affinity partitioning. A method for purification of proteins using specific polymer-ligands in aqueous two phase systems. *J. Biol. Chem. 250*: 1484-1489.

Hubert, P., Dellacherie, E., Neel, J., and Baulieu, E.-E. (1976). Affinity partitioning of steroid binding proteins. The use of polyethylene oxide bound estradiol for purifying Δ_{5-4} 3-oxosteroid isomerase. *FEBS Lett. 65*: 169-174.

Johansson, G. (1976). The effect of poly(ethylene glycol) esters on the partition of proteins and fragmented membranes in aqueous biphasic systems. *Biochim. Biophys. Acta 451*: 517-529.

Johansson, G., Gysin, R., and Flanagan, S. D. (1981). Affinity partitioning of membranes: Evidence for discrete membrane containing cholinergic receptor. *J. Biol. Chem. 256*: 9126-9135.

Kopperschläger, G., and Johanssen, G. (1982). Affinity partitioning with polymer bound Cibacron Blue F3G-A for rapid large-scale purification of phosphofructokinase from baker's yeast. *Anal. Biochem.* *124*: 117-124.

Kopperschläger, G., Böhme, H. J., and Hofmann, E. (1982). Cibacron Blue F3G-A and related dyes as ligands in affinity chromatography. In *Advances in Biochemical Engineering*. T. K. Ghose, A. Fiechter, and N. Blakebrough (eds.). Springer-Verlag, Berlin, pp. 101-138.

Kopperschläger, G., Lorenz, G., and Usbeck, E. (1983). Application of affinity partitioning in an aqueous two-phase system to the investigation of triazine dye-enzyme interactions. *J. Chromatogr.* *259*: 97-105.

Kroner, H. K., Cordes, A., Schelper, A., Morr, M., Bückmann, A. F., and Kula, M.-R. (1982). Affinity partition studied with glucose-6-phosphate dehydrogenase in aqueous two-phase system in response to triazine dyes. In *Affinity Chromatography and Related Techniques*, T. C. J. Gribnau, J. Visser, and R. J. F. Nivard (eds.). Elsevier Scientific, Amsterdam, pp. 491-501.

Kula, M.-R., Johansson, G., and Bückmann, A. F. (1979). Large-scale isolation of enzymes. *Biochem. Soc. Trans.* *7*: 1-6.

Lowe, C. R., Clonis, Y. D., Goldfinch, M. J., Small, D. A. P., and Atkinson, A. (1982). Some preparative and analytical applications of triazine dyes. In *Affinity Chromatography and Related Techniques*, T. C. J. Gribnau, J. Visser, and R. J. F. Nivard (eds.). Elsevier Scientific, Amsterdam, pp. 389-398.

Shanbhag, V. P., and Johansson, G. (1974). Specific extraction of human serum albumin by partition in aqueous biphasic systems containing poly(ethylene glycol) bound ligands. *Biochem. Biophys. Res. Commun* *61*: 1141-1146.

18
Affinity Electrophoresis

WERNER SCHÖSSLER
Central Institute of Cardiovascular Research
of the Academy of Sciences of the GDR

The definition of affinity electrophoresis is, by analogy with affinity chromatography, the electrophoresis of a charged molecule in an electrical field within a carrier containing immobilized specific interaction partner(s). Obviously, the principle is quite similar to immunoelectrophoresis methods, such as crossed immunoelectrophoresis and rocket immunoelectrophoresis, but these techniques are not covered here.

The first experiments of specific interaction of a gel carrier with an electrophoresed protein were reported by Entlicher and colleagues (1969), as well as Takeo and Nakamura (1972). But the term "affinity electrophoresis" was introduced later by Bøg-Hansen (1973). In general, two different techniques in affinity electrophoresis are used: (a) affinity electrophoresis of glycoproteins in agarose gels containing lectins and detection of the glycoprotein with a specific antiserum in the second dimension (Bøg-Hansen, 1973; Bøg-Hansen and Brogren, 1975), and (b) affinity electrophoresis in polyacrylamide gel containing immobilized ligands interacting with (glyco)protein-lectin complexes (Hořejši and Kocourek, 1974a).

AFFINITY ELECTROPHORESIS IN AGAROSE GELS

The basic principle of the agarose gel method according to Bøg-Hansen is a combination of affinity electrophoresis with lectins and quantitative immunoelectrophoresis. The approaches developed by Bøg-Hansen and coworkers in this electrophoretic system are that the interaction

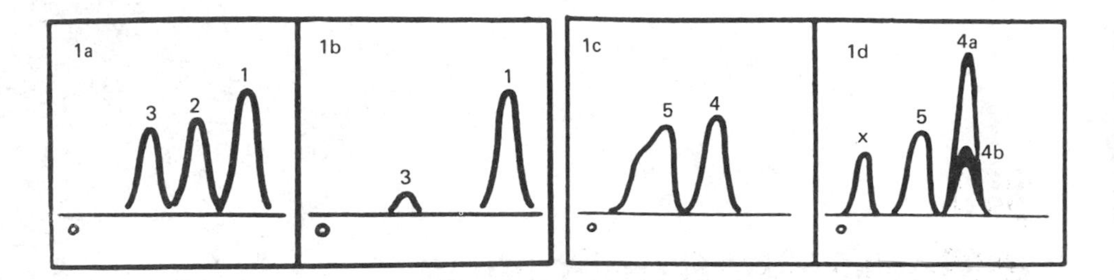

FIGURE 18.1 Characteristic precipitation patterns of glycoproteins after reaction with immobilized lectin before crossed immunoelectrophoresis. (a) and (c) Control experiments without lectin. (b) and (d) The pattern after reaction with lectin. (c) and (d) Reaction of complex glycoproteins. (From Bøg-Hansen, 1979a.)

between the lectin and the corresponding glycoprotein can take place before, during or after electrophoresis (Bøg-Hansen, 1979a, 1983).

Reaction with Immobilized Lectin Before Electrophoresis

By this method immobilized lectin is mixed with the protein sample, and after centrifugation the supernatant is analyzed by crossed immunoelectrophoresis. The following precipitation patterns can be expected (Fig. 18.1). In this schematic representation protein 1 has no affinity to the lectin and its precipitate is unchanged and may be used as an internal reference in comparison to the other protein. Typical results are disappearance (protein 2) or the reduction in the area of the precipitate (protein 3). In special cases the precipitate could split into several precipitates (protein 4 → 4a + 4b) or in precipitates with different first-dimension mobilities (protein 5). This method included some technical problems (only with small amounts of lectins is the precipitation pattern affected by the lectin concentration) and supplied enough information about the interaction partners.

Interaction with an Immobilized Lectin in an Intermediate Gel

This method was introduced by Bøg-Hansen (1973) under the term "crossed immuno-affinoelectrophoresis" and is comparable with the use of specific antibodies in an intermediate gel for identification of antigens. There are some possibilities of reactions that can be given the following characteristic precipitation patterns corresponding to Bøg-Hansen (1979a) (Fig. 18.2). Typical reactions are the disappearance (protein 2) or the reduction in the area of the precipitate

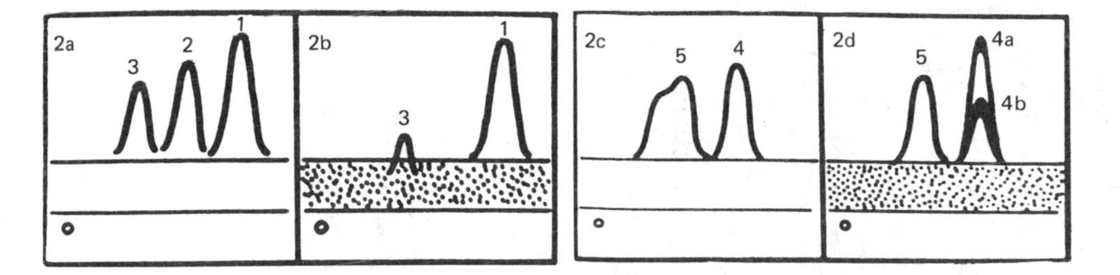

FIGURE 18.2 Characteristic precipitation patterns of glycoproteins after electrophoresis in the first dimension and interaction with the lectin in an intermediate gel. (a and c) The control experiment without lectin. (b and d) The pattern after reaction with lectin. (From Bøg-Hansen, 1979a.)

(protein 3). Similar to the first method, the splitting of precipitation profile (protein 5) can be seen. A list of further observed precipitation lines, including experiments with free lectins (the pH of the buffer used and the electroendosmosis of agarose must be chosen so that the free lectin incorporated into the agarose is immobile), was given by Bøg-Hansen (1979b). Since its introduction, the method has been used for many glycoproteins and for the prediction of separation experiments (Bøg-Hansen, 1973; Bøg-Hansen and Brogren, 1975; Bøg-Hansen et al., 1978; Schmidt-Ullrich et al., 1975; Bjerrum and Bøg-Hansen, 1976; Bisati et al., 1979; Gerlach et al., 1979).

Lectin in the First-Dimension Gel

The principle of incorporating lectin in the first-dimension gel was introduced by Bøg-Hansen and Brogren (1975) and is analogous to the well-known use of specific antibodies in the first-dimension gel for detection and identification of antigens in crossed immunoelectrophoresis. This method has been adopted by many authors, because in this way more information about the interaction partners can be obtained. Figure 18.3 shows the most typical precipitation pattern when glycoproteins are electrophoresed through a gel containing immobilized or free lectin. The pattern is highly dependent upon the lectin concentration. Similar to the first two methods, protein 1 is a protein that does not have an interaction with the lectin and is taken again as an internal reference. Other typical reactions are the disappearance of precipitate (protein 2) or the reduction in the area and/or changed position and changed precipitation profiles (protein 3).

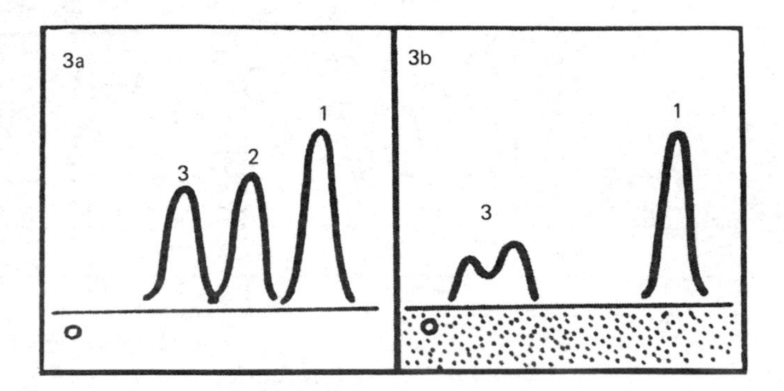

FIGURE 18.3 Characteristic precipitation patterns after reaction with lectin in the first-dimension gel. (a) The control experiment without lectin. (b) The pattern after interaction with lectin. (From Bøg-Hansen, 1979a.)

As mentioned, this method has been used by many authors for several purposes, such as for the identification and characterization of glycoproteins and prediction of separation experiments (Bøg-Hansen and Brogren, 1975; Bøg-Hansen et al., 1978; Bjerrum and Bøg-Hansen, 1976; Nielsen and Bjerrum, 1977; Fournier et al., 1979; Guldager and Bøg-Hansen, 1979; Kerckaert et al., 1979; Nilsson and Bøg-Hansen, 1979; Ory et al., 1979).

The main advantages of the agarose gel techniques, especially of the third method are as follows.

The method provides a means for specific identification of the components of a protein mixture.

Therefore, it is not necessary to purify the components.

The method reveals the presence of different molecular species of a glycoprotein.

The method can be used for studies of interaction for other macromolecules.

Apart from this, the use of antibodies limits the analytic use of lectins in some cases because of the glycoprotein nature of the antibodies and the necessity for nondenaturating conditions during electrophoresis. Additionally, the resolution power of agarose gels is lower than that of polyacrylamide gels.

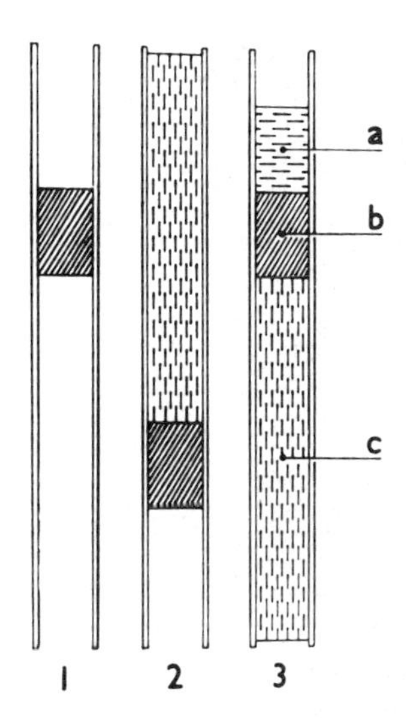

FIGURE 18.4 Procedure of gel setting for affinity electrophoresis: (a) large-pore gel, (b) affinity gel, (c) small-pore gel. (From Hořejši and Kocourek, 1974a.)

AFFINITY ELECTROPHORESIS IN POLYACRYLAMIDE GELS

Perhaps this was the reason Hořejši and Kocourek (1974a) developed the polyacrylamide technique. Originally they used the conventional polyacrylamide electrophoresis in this manner, so that the common polymerization mixture of acrylamide and bisacrylamide was supplement with allylglycosides as additional monomers. The resulting gels contain immobilized sugars, which are in general very potent materials for affinity chromatography of lectins, too (Hořejši and Kocourek, 1974b). The principle of the method, usually performed in tubes, is illustrated in Figure 18.4 (Hořejši and Kocourek, 1974a).

At first the affinity gel is produced by copolymerization of acrylamide, bisacrylamide, and the corresponding allylglycosides. After the polymerization reaction, the gel rods must be removed from glass tubes, washed free from the unreacted allylglycosides, and equilibrated with the buffer and then inserted back into glass tubes. In the second step the tube is turned around and the small gel polymerization is done (usually 5–7.5%). Last, the tube is returned and the large-pore gel (3%) is polymerized. This laborious and time-consuming procedure is the "Achilles heel" of the method. Therefore, Hořejši and coworkers improved the preparation of affinity gels by entrapment of suitable macromolecules (dextran) containing covalently bound sugar residues into the polyacrylamide gel network (Čeřovsky et al., 1980) (a review of this and other methods was given by Hořejši, 1981). The danger of this method and the method described by Hořejši and coworkers (1978) is that the microdistribution of the ligand in affinity gel is nonhomogeneous, which will influence the effective concentration of the immobilized ligand. Recently, Hořejši and colleagues (1982a) described a new method for the preparation of affinity gels by coupling to periodate-treated agarose beads, which is similar to a method developed in our laboratory (Schössler and Dittrich, 1982). By this method homogeneous affinity gels are obtained by mixing the substituted beads with melted agarose or with the polymerization mixture prepared for polyacrylamide gels.

The possibilities of the polyacrylamide gel method in affinity electrophoresis have been excellently represented by Hořejši and coworkers (1979). This overview is shown in Figure 18.5. In the case a affinity electrophoresis is used for the identification of a lectin interacting with affinity gel only for qualitative applications. The mobility of a lectin depends on the concentration of immobilized and free sugar in the gel (Fig. 18.5). In case b the concentration of the sugar in gel 2 is lower than in gel 3. When affinity electrophoresis is performed in gels containing a constant concentration of immobilized sugar but different concentrations of free sugar, the mobility of the lectin increases (gels 4 and 5). Similar to the agarose gel method, the polyacrylamide gel method offers the possibility to distinguish between different molecular species (case c in Fig. 18.5). Very illustrative are the examples for the use of affinity electrophoresis in monitoring the results of chemical modification of proteins given by Hořejši and colleagues (1977b, 1979). If a modification reaction results in a 90% loss of specific activity, this can be caused by inactivation of 90% and the remaining 10% are fully active (gel 4), or by inactivation of all molecules to 10% (gel 5), or by splitting off the protein in a complex mixture of reaction products (gel 6).

QUANTITATIVE AFFINITY ELECTROPHORESIS

It is well established that affinity chromatography is a valuable tool for the determination of affinity constants from the interaction between macromolecules and high- or low-molecular-weight ligands. The dependence of the electrophoretic mobility of a protein from the concentration of the immobilized or free ligand can be used for the determination of the dissociation constants of the protein-immobilized ligand complex and the protein-free ligand complex, respectively.

The magnitude of the electrophoretic shift caused by the interaction with the ligand is an expression of the affinity between the interacting partners. The simplest way to express the relative affinity of a protein to an interacting ligand is the calculation of the retardation coefficient (Bøg-Hansen et al., 1978):

$$R = \frac{d_0}{d - 1} \tag{1}$$

where d and d_0 are the migration distances with and without ligand, respectively. The advantage of this method is that, for the determination of retardation coefficient, only one experiment must be performed with the ligand in the gel. On the other hand, the retardation coefficient is a relative quantity and not comparable to other methods.

The dependence of electrophoretic mobility of the protein on both the concentration of immobilized and free ligand can be used for the calculation of the dissociation constants by means of a special plot (Takeo and Kabat, 1978).

Hořejši (1979a, 1979b), Hořejši and Tichá (1981a), and Bøg-Hansen and Takeo (1980), among others, have developed generally applicable equations for the determination of dissociation constants. The relation of the relative migration mobility of the protein and the ligand concentration is

$$\frac{d}{d_0 - d} = \frac{K_i}{c_i}\left(1 + \frac{c}{K}\right) \tag{2}$$

where K_i is the dissociation constant of the protein-immobilized ligand complex, K is the dissociation constant of the protein-free ligand complex, c_i and c are the concentrations of the immobilized or free ligand in affinity gel, respectively, and d and d_0 are the mobilities of the protein in the affinity gel and the control experiment without the ligand.

When plotting the dependence of $d/d_0 - d$ against c, a straight line is obtained that intercepts the y axis at the point equal to K_i/c_i, and the intercept on the x axis is -K. In the simple case where the concentration of the free ligand is zero (c = 0), Equation (2) becomes Equation (3):

$$\frac{1}{d_0 - d} = \frac{K_i}{d_0 c_i} + \frac{1}{d_0} \tag{3}$$

By plotting $1/(d_0 - d)$ against $1/c_i$, the dissociation constant K_i can be calculated from the intercept with the x axis ($-1/K_i = 1/c_i$). However,

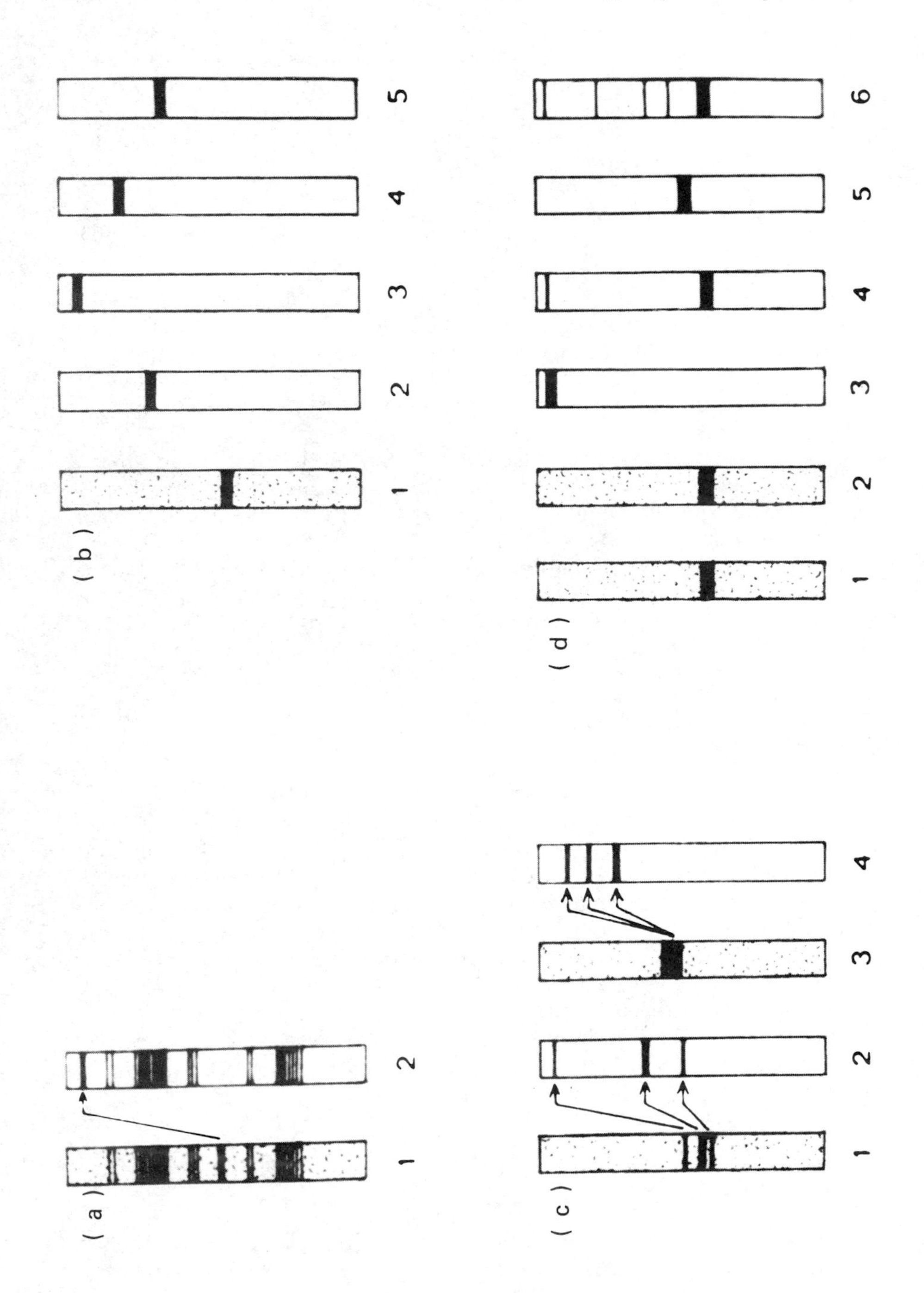
(a)
1
2
(b)
1
2
3
4
5
(c)
1
2
3
4
(d)
1
2
3
4
5
6

FIGURE 18.5 Schematic representation of reactions in affinity electrophoresis in polyacrylamide gel. (a) Identification of a component interacting with affinity gel (gel 2). (b) Dependence of mobility of concentration of immobilized (c_i) and free ligands (c) in an affinity gel. Gels 2 and 3 contain immobilized ligand ($c_{i3} > c_4$); gels 4 and 5 contain constant concentration of immobilized ligand and variable amounts of free ligand ($c_5 > c_4$). (c) Presence of substances with different binding affinity toward immobilized ligand (gel 2); resolution of a protein into molecular forms differing in affinity toward immobilized ligand (gel 4). (d) The use of affinity electrophoresis for monitoring of binding properties of chemically modified proteins. (1) Native protein, (2) modified protein with a 90% loss of activity), (3) native protein on affinity gel, (4-5) different possibilities of precipitation patterns. Area of control gel without immobilized ligand is stippled. (From Hořejši et al., 1979.)

these simple relations are exactly valid only under several assumptions, for example, that the complex formation is very rapid, the microdistribution of the ligand is homogeneous, the protein has only a single binding site, and the protein concentration is lower than the ligand concentration. Hořejši (1979a, 1979b) and Hořejši and Tichá (1981a) have developed equations accounting for the multivalency of proteins or special cases, such as for the determination of the effective ligand concentration and the protein concentration. For details the reader is referred to the original literature.

For proteins that have good electrophoretic properties no simple alternative method for the determination of the affinity between macromolecules and high- or low-molecular-weight ligands has been published up to now. A prerequisite for the exact determination of the dissociation constants K_i is the accuracy of the determination of the immobilized ligand concentration (which can be difficult) and that the mobility of the complex is not influenced by the gel matrix. Therefore, agarose gel modification is preferred in investigating macromolecule-macromolecule interactions because the pores allow free penetration of macromolecules up to 2×10^7 daltons (1% agarose). The polyacrylamide modification is especially useful for studying macromolecule-small ligand interactions.

The methods described here for the determination of dissociation constants have been used by many authors for determination of dissociation constants of complexes between lectins and sugars (Hořejši et al., 1977a), enzymes and their substrates (Bøg-Hansen, 1979a), enzymes and Blue Dextran (Tichá et al., 1978), and lectins and glycoproteins (Bøg-Hansen, 1973, 1983; Bøg-Hansen and Brogren, 1975), as well as mouse antidextran myeloma proteins with isomaltose oligosaccharide (Takeo and Kabat, 1978).

Despite its limitations, affinity electrophoresis has some advantages for analytic purposes.

The method can be used for studies of interacting macromolecules.
A very small sample of material is needed.
It is not necessary to purify interacting components.
A multitude of proteins reacting with the ligand may be studied simultaneously.
The dissociation constants of different molecules or different molecular species are determined simultaneously.
Dissociation constants in the range from 10^2 to 10^{-7} M can be studied
It is possible to measure dissociation constants in dependence on pH, temperature, urea, and other factors.
The method is simple and can be performed with ordinary laboratory equipment.
The method can be used for prediction of preparative experiments.

CORRELATION OF AFFINITY CHROMATOGRAPHY AND AFFINITY ELECTROPHORESIS

Since the initial experiments by Bøg-Hansen (1973), several authors (Bøg-Hansen et al., 1978; Nilsson and Bøg-Hansen, 1979) have used affinity electrophoresis for predicting the results of affinity chromatography, especially in cases where material is scarce. In a very impressive experiment studying α-fetoprotein (AFP), Kerckaert and coworkers (1979) have found a strong correlation between affinity electrophoresis and affinity chromatography (Fig. 18.6). They have found a nonreactive (a) fraction to ConA, a weakly ConA-reactive (b), and a ConA-reactive fraction (eluted by α-methyl-D-glucopyranoside) in affinity chromatography.

From this and similar experiments (Bøg-Hansen, 1973), it has often been concluded that affinity electrophoresis is the method of choice to predict fractionation experiments by means of column technique. In a critical study, Ramlau and Bock (1979) have shown that affinity electrophoresis has some limitations in the prediction of chromatography experiments because no information is given about the capacity of the matrix, the elution conditions, or recovery.

APPLICATIONS

Affinity electrophoresis has been developed into an established method in the research laboratory as well as clinical laboratory in recent years. The potential possibilities of affinity electrophoresis are shown by a few examples, but this choice does not attempt to be complete (Table 18.1).

The interactions of glycoproteins and lectins are investigated above all by Bøg-Hansen and coworkers, whereas Hořejši and coworkers have extensively studied the interaction between lectins and sugars. An overview of the different modifications of affinity electrophoresis and their applications was given by Bøg-Hansen and colleagues (1977).

As mentioned, one of the advantages of affinity electrophoresis is that only a small amount of material is required, and therefore, it can be used routinely for diagnosis in the clinical laboratory. By performing the crossed immunoaffinoelectrophoresis the pattern of ConA variants of α-fetoprotein in amniotic fluid seems to be valuable in the diagnosis of fetal abnormalities (Smith et al., 1979).

Affinity electrophoresis has also been used in the differentiation of fetal and hepatome α-fetoprotein lectin-bound fractions (Kerckaert et al., 1979; Smith and Kelleher, 1980).

Some new clinical applications of affinity electrophoresis were represented recently (Hirai, 1984). The application of affinity electrophoresis in the variant as crossed immunoaffinoelectrophoresis

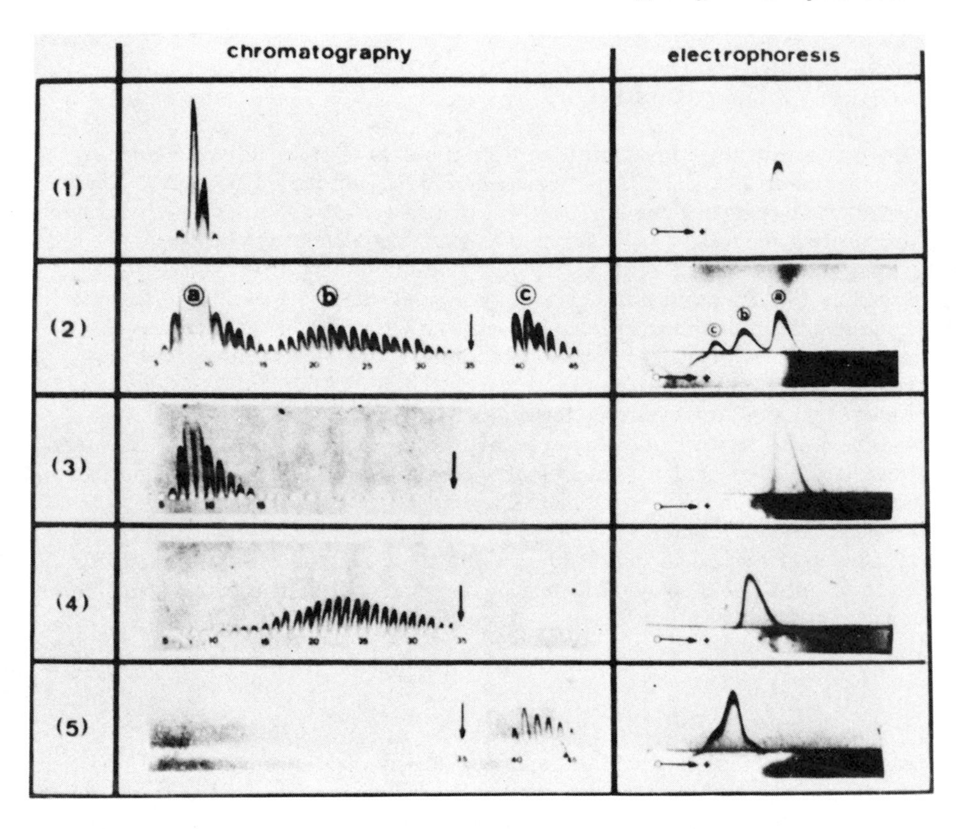

FIGURE 18.6 Comparison between affinity chromatography and affinity electrophoresis of rat α-fetoprotein. (1) Control experiment (Sepharose column or agarose gel in affinity electrophoresis). (2) Total rat α-fetoprotein patterns (α-fetoprotein from chromatography was determined by rocket immunoelectrophoresis). The three isolated fractions were then submitted again to ConA affinity chromatography and affinity electrophoresis. (3) ConA nonreactive fraction. (4) ConA weakly reactive fraction. (5) ConA reactive fraction (the arrow indicates the elution with 0.1 M α-methyl-D-glucopyranoside). (From Kerckaert et al., 1979.)

according to Bøg-Hansen (1973), or as polyacrylamide affinity electrophoresis according to Hořejši and Kocourek (1974a), has been essentially focused to the study of interactions between lectins and glycoproteins, as well as lectins and sugars or enzymes and the corresponding substrates, respectively.

Affinity electrophoresis is widely suitable to study macromolecule-macromolecule interactions, protein-nucleic acid interactions, and interactions between hormones and receptors. We have recently adapted the method of Bøg-Hansen for studying the interactions of factor VIII-related antigen with vessel wall proteins (Schössler and Dittrich, 1982) and used the term "crossed immunoaffinity electrophoresis." The background of these studies is the well-known fact that platelet adhesion to vascular subendothelium is considered to be an important step in thrombosis and atherogenesis. Factor VIII-related antigen (F VIII R:Ag) can play an important role in these interactions. Figure 18.7 shows the crossed immunoaffinity electrophoresis of F VIII R:Ag bound to collagen. The dependence of the electrophoretic mobility of F VIII R:Ag from the collagen concentration is shown in Figure 18.8. By using Equation (3) the dissociation constant of the collagen F VIII R:Ag interaction can be determined with 2×10^{-7} M. This strong interaction is specific because no binding of F VIII R:Ag to elastin could be found (unpublished results), and it may play a key role in the adhesion process of platelets.

FUTURE DEVELOPMENTS

Although affinity electrophoresis has been widely accepted and used in a large number of studies, the method has some problems. Thus, it is necessary to develop fast and simple procedures to determine the effective concentration of immobilized ligand in affinity gel. This is important for the exact determination of dissociation constants.

The principle of affinity electrophoresis can be extended to other electrophoresis or histochemical techniques. Thus, the use of affinity isoelectric focusing (Hořejši and Tichá, 1981b) or affinity isotachophoresis (Hořejši et al., 1982b) may be important in the future. Moreover, the use of different matrices in affinity chromatography as hydrophobic materials (Chen and Moravetz, 1981) or group-specific adsorbents will extend the potential of this method. Application of affinity electrophoresis for preparative purposes has not been reported up to now. This modification would be advantageous in isolation of proteins under nondenaturating conditions. Apart from the electrophoretic methods described, the binding of labeled lectins or specific antibodies (labeled by radioactive markers, enzymes, fluorochromes, and others) for identification and quantification of (glyco)-proteins by using conventional histochemical staining methods or autoradiography after electrophoresis could be important.

TABLE 18.1 Various Modifications of Affinity Electrophoresis with Interacting Components and Some Applications

Modification	Interacting component electrophoresed	Interacting component in the medium	Purpose
Crossing diagrams	Enzyme	Substrate inhibitor	Identification of interacting components
	Glycoproteins	Lectins	
Affinity electrophoresis	Lectin	Carbohydrate	Identification of lectins
	Enzyme	Substrates	Determination of dissociation constants
One-dimensional affinity electrophoresis	Glycoprotein enzymes	Immobilized lectin	Identification of glycoprotein enzymes
Rocket-affinity electrophoresis	Glycoproteins	Lectin	Quantification of glycoproteins
	Lectin	Glycoproteins	Quantification of lectins
Fused rocket-affinity electrophoresis	Glycoproteins	Lectins	Analysis of Fractionations Progressive changes Treatments

Crossed affinity electrophoresis	Lectin	Glycoproteins	Identification of lectins
	Glycoproteins	Lectin	Identification and quantification of lectin-binding glycoproteins
Crossed immunoelectrophoresis with ligand in first dimension	Glycoproteins	Lectin	Identification of interacting components
	Lectins	Glycoproteins	Determination of binding specificity
			Analysis of microheterogeneity
			Determination of dissociation constants
Crossed immunoelectrophoresis with ligand in an intermediate gel	Glycoproteins	Lectins	Identification of interacting components
	Lectins	Glycoproteins	Partial characterization of (number of binding sites)

Source: From Bøg-Hansen et al. (1977).

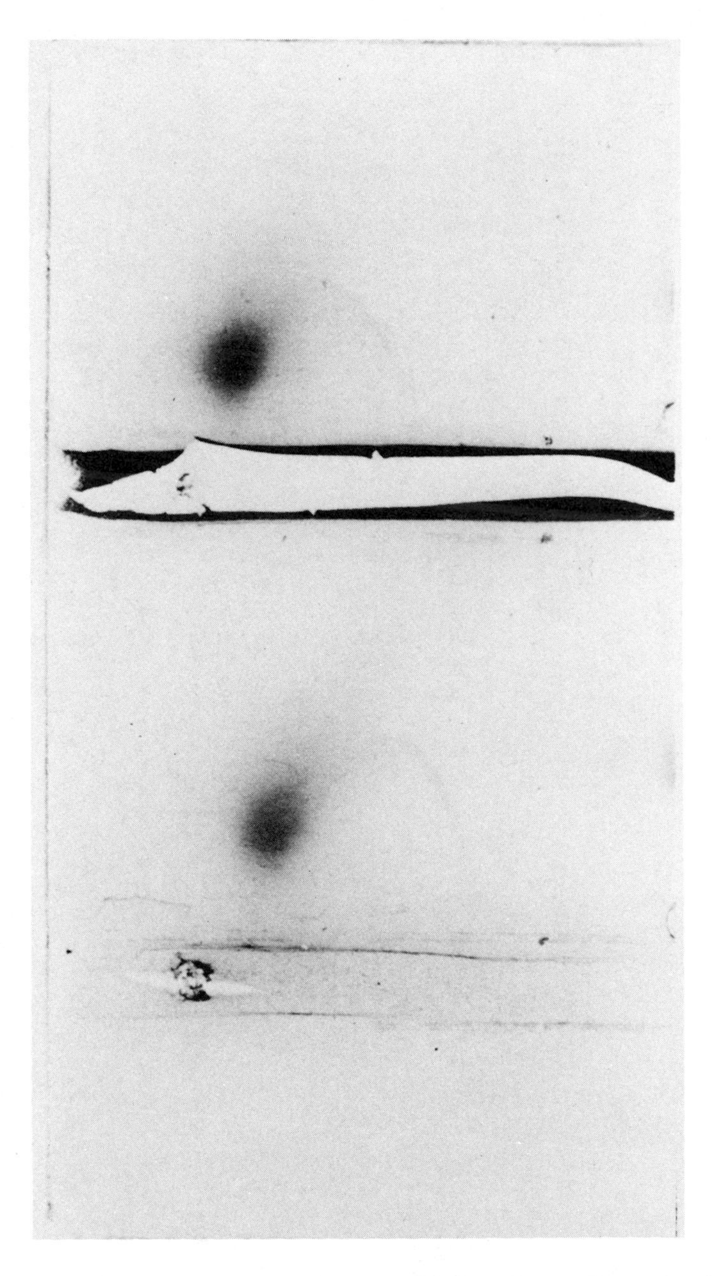

FIGURE 18.7 Crossed immunoaffinity electrophoresis of F VIII R:Ag. The first dimension contains collagen Sepharose (1 mg collagen per milliliter gel) (top) or in the control experiment agarose (bottom; the second dimension contains 0.5% antiserum to human factor VIII) 0 = start hole.

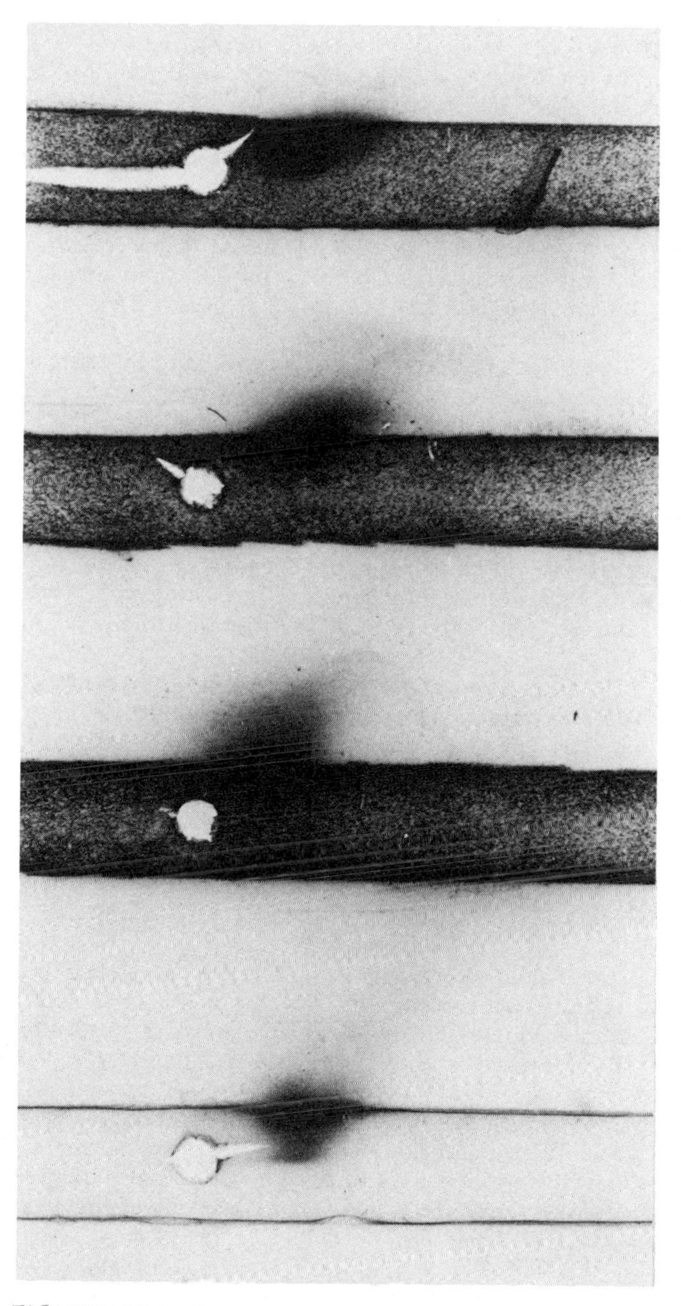

FIGURE 18.8 Dependence of the electrophoretic mobility of F VIII R: Ag from the collagen concentration 0, 41, 28, and 21 μg/ml from bottom to top).

In conclusion, affinity electrophoresis has been developed into an established method, but much remains to be done to extend the applications of this method in the next few years.

REFERENCES

Bisati, S., Brogren, C. H., and Simonsen, M. (1979). Major lectin binding plasma membrane proteins and histocompatibility antigens studied by quantitative immunoelectrophoresis. *Protides Biol. Fluids 27*: 471-474.

Bjerrum, O. J., and Bøg-Hansen, T. C. (1976). The immunochemical approach to the characterization of membrane proteins. Human erythrocyte membrane proteins analyzed as a model system. *Biochim. Biophys. Acta 455*: 66-89.

Bøg-Hansen, T. C. (1973). Crossed immuno-affinoelectrophoresis – an analytical method to predict the result of affinity chromatography. *Anal. Biochem. 56*: 480-488.

Bøg-Hansen, T. C. (1979a). Affinity electrophoresis with lectins for characterization of glycoproteins. In *Affinity Chromatography and Molecular Interactions*, J.-M. Egly (ed.). Editions INSERM, Paris, pp. 399-416.

Bøg-Hansen, T. C. (1979b). Some characteristic reactions between glycoproteins and lectins in analytical affinity electrophoresis. *Protides Biol. Fluids 27*: 659-664.

Bøg-Hansen, T. C. (1983). Affinity electrophoresis of glycoproteins. In *Solid Phase Biochemistry: Analytical and Synthetic Aspects*, W. H. Scouten (ed.). John Wiley & Sons, New York.

Bøg-Hansen, T. C., and Brogren, C. H. (1975). Identification of glycoproteins with one and with two or more binding sites to con A by crossed immuno-affinoelectrophoresis. *Scand. J. Immunol. 4, Suppl. 2*: 135-139.

Bøg-Hansen, T. C., and Takeo, K. (1980). Determination of dissociation constants by affinity electrophoresis: Complexes between human serum proteins and concanavalin A. *Electrophoresis 1*: 67-71.

Bøg-Hansen, T. C., Bjerrum, O. J., and Brogren, C.-H. (1977). Identification and quantification of glycoproteins by affinity electrophoresis. *Anal. Biochem. 81*: 78-87.

Bøg-Hansen, T. C., Prahl, P., and Løwenstein, H. (1978). A set of analytical electrophoresis experiments to predict the results of affinity chromatographic separations: Fractionation of allergens from cow's hair and dander. *J. Immunol. Methods 22*: 293-307.

Čeřovsky, V., Tichá, M., Hořejši, V., and Kocourek, J. (1980). Studies on lectins:XLIX. The use of glycosyl derivates of dextran T 500 for affinity electrophoresis of lectins. *J. Biochem. Biophys. Methods 3*: 163-172.

Chen, J.-L., and Moravetz, H. (1981). Affinity electrophoresis in gels containing hydrophobic substituents. *J. Biol. Chem. 256*: 9221-9223.

Entlicher, G., Tichá, M., Koštiř, J. V., and Kocourek, J. (1969). Studies on phythemagglutinins. II. Phythemagglutinins of Pisum sativum L. and Lensescutenta moench: Specific interactions with carbonhydrates. *Experientia 25*: 17-18.

Fournier, C., Kerckaert, J.-P., Bayard, B., Colly, M., and Biserte, B. (1979). Crossed lectin-immunoelectrophoresis of in vivo and in vitro produced rat alphafetoprotein. *Protides Biol. Fluids 27*: 623-626.

Gerlach, J., Bjerrum, O. J., Rank, H. C., and Bøg-Hansen, T. C. (1979). Crossed immunoelectrophoresis of yeast plasma membrane lectine receptors. *Protides Biol. Fluids 27*: 479-482.

Guldager, P., and Bøg-Hansen, T. C. (1979). Pea seed lectin (PEA). *Protides Biol. Fluids* 27:401-404.

Hirai, H. (ed.) (1984). *Electrophoresis 83.* Walter de Gruyter, Berlin.

Hořejši, V. (1979a). Some theoretical aspects of affinity electrophoresis. *J. Chromatogr. 178*: 1-13.

Hořejši, V. (1979b). Affinity electrophoresis – theory and some applications. In *Affinity Chromatography and Molecular Interactions*, J.-M. Egly (ed.). Editions INSERM, Paris, pp. 391-398.

Hořejši, V. (1981). Affinity electrophoresis. *Anal. Biochem. 112*: 1-8.

Hořejši, V., and Kocourek, J. (1974a). Studies on phythemagglutinins. XVIII. Affinity electrophoresis of phythemagglutinins. *Biochim. Biophys. Acta 336*: 338-343.

Hořejši, V., and Kocourek, J. (1974b). Affinity electrophoresis: Separation of phythemagglutinin on o-glycosyl polyacrylamide gels. *Methods Enzymol. 34*: 361-367.

Hořejši, V., and Tichá, M. (1981a). Theory of affinity electrophoresis. Evaluation of the effects of protein multivalency, immobilized ligand heterogeneity and microdistribution and determination of effective concentration of immobilized ligand. *J. Chromatogr. 216*: 43-62.

Hořejši, V., and Tichá, M. (1981b). Affinity isoelectric focusing in polyacrylamide gel – a method to detect ligand-binding proteins. *Anal. Biochem. 116*: 22-26.

Hořejši, V., Tichá, M., and Kocourek, J. (1977a). Studies on lectins. XXXI. Determination of dissociation constants of lectin-sugar complexes by means of affinity electrophoresis. *Biochim. Biophys. Acta 499*: 290-300.

Hořejši, V., Tichá, M., and Kocourek, J. (1977b). Studies on lectins. XXXII. Application of affinity electrophoresis to the study of the interaction of lectins and their derivatives with sugars. *Biochim. Biophys. Acta 499*: 301-308.

Hořejší, V., Smolek, P., and Kocourek, J. (1978). Studies on lectins. XXXV. Water-soluble o-glycosyl polyacrylamide derivatives for specific preparation of lectins. *Biochim. Biophys Acta 538*: 293-298.

Hořejší, V., Tichá, M., and Kocourek, J. (1979). Affinity electrophoresis. *Trends Biochem. Sci. 4*: N6-N7.

Hořejší, V., Tichá, M., Tichý, P., and Holý, A. (1982a). Affinity electrophoresis: New simple and general methods of preparation of affinity gels. *Anal. Biochem. 125*: 358-369.

Hořejší, V., Datta, T. K., and Tichá, M. (1982b). Affinity electrophoresis in an isotachophoretic discontinuous buffer system. *J. Chromatogr. 241*: 395-398.

Kerckaert, J. P., Bayard, B., and Biserte, G. (1979). Microheterogeneity of rat, mouse, and human α_1-fetoprotein as revealed by polyacrylamide gel electrophoresis and by crossed immuno-affino-electrophoresis with different lectins. *Biochim. Biophys. Acta 576*: 99-108.

Nielsen, C. S., and Bjerrum, O. J. (1977). Crossed immunoelectrophoresis of bovine milkfat globule membrane protein solubilized with non-ionic detergent. *Biochim. Biophys. Acta 466*: 496-509.

Nilsson, M., and Bøg-Hansen, T. C. (1979). An approach to fractionation of human proteins with some lectins. *Protides Biol. Fluids. 27*: 599-602.

Ory, R. L., Mod, R., and Bog-Hansen, T. C. (1979). Some properties of ricegerm agglutinin (RGA). *Protides Biol. Fluids 27*: 387-390.

Ramlau, J., and Bock, E. (1979). Affinity interaction immuno-electrophoresis for characterization and prediction of fractionation of antigens. In *Affinity Chromatography and Molecular Interactions*, J.-M. Egly (ed.). Editions INSERM, Paris, pp. 147-174.

Schmidt-Ullrich, R., Wallach, D. F. H., and Hendricks, J. (1975). Concanavalin A – reactive protein of rabbit thymocyte plasma membranes: Analysis by crossed immune electrophoresis and sodium dodecyl sulfate/polyacrylamide gel electrophoresis. *Biochim. Biophys. Acta 382*: 295-310.

Schössler, W., and Dittrich, C. (1982). Crossed immuno-affinity electrophoresis -- a method for analysing the interactions of factor VIII-related antigen with vessel wall proteins. *Thromb. Res. 28*: 677-680.

Smith, J. C., and Kelleher, P. C. (1980). Alpha-fetoprotein molecular heterogeneity. Physiologic correlations with normal growth, carcinogenesis and tumor growth. *Biochim. Biophys. Acta 605*: 1-32.

Smith, J. C., Kelleher, P. C., Belanger, L., and Dallaire, L. (1979). Reactivity of amniotic fluid alpha-fetoprotein with concanavalin A in diagnosis of neutral tube defects. *Brit. Med. J. 1*: 920-921.

Takeo, K., and Kabat, A. (1978). Binding constants of dextrans and isomaltose oligosaccharides to dextran specific myeloma proteins determined by affinity electrophoresis. *J. Immunol. 121*: 2305-2310.
Takeo, K., and Nakamura, S. (1972). Dissociation constants of glucan phosphorylase of rabbit tissues studied by polyacrylamide gel disc electrophoresis. *Arch. Biochem. Biophys. 153*: 1-7.
Tichá, M., Hořejši, V., and Barthová, J. (1978). Affinity electrophoresis of proteins interaction with blue dextran. *Biochim. Biophys. Acta 534*: 58-63.

19
High Performance Liquid Affinity Chromatography

High-performance liquid affinity chromatography (HPLAC) combines the pronounced adsorption specifity of affinity matrices with the benefits of high-pressure liquid chromatography (HPLC), such as their rapid practicability and the high resolution of the elution diagrams. This method was introduced into practice first by Ohlson and colleagues (1978) using modified silica or glass by reason of (a) its chemical and mechanical stability, withstanding the high-pressure drops, and (b) its high porosity, allowing free penetration of high-molecular-weight compounds (c) Moreover, it can be easily derivatized and activated for ligand introduction (Chaps. 4 and 5). To exclude or to minimize nonspecific adsorption effects, a hydrophobic layer of glyceropropyl groups can be covalently attached, for example, by silanization with glyceropropylsilane to diol-silica (Regnier and Noel, 1976). By selection of suitable silanes, active groups for the attachment of specific ligands can also be introduced, such as epoxy groups, when epoxy silane as a hydrophobic layer-forming material is used (Fig. 19.1). Another way of activating hydrophilized silica is the chemical modification of groups covering the silica particles, such as the formation of aldehyde silica by periodate oxidation of diol silica, as illustrated in Figure 19.1. Applying such adsorbent materials the essential varieties of common biospecific affinity chromatography have been introduced in the last few years in HPLAC, as shown in Table 19.1 (Mosbach et al., 1982).

As an example demonstrating the advantage of HPLAC compared with the common affinity chromatography, in Figure 19.2 (Larsson et al., 1979) the separation of liver alcohol dehydrogenase and lactate dehydrogenase is shown. Both enzymes are adsorbed by the

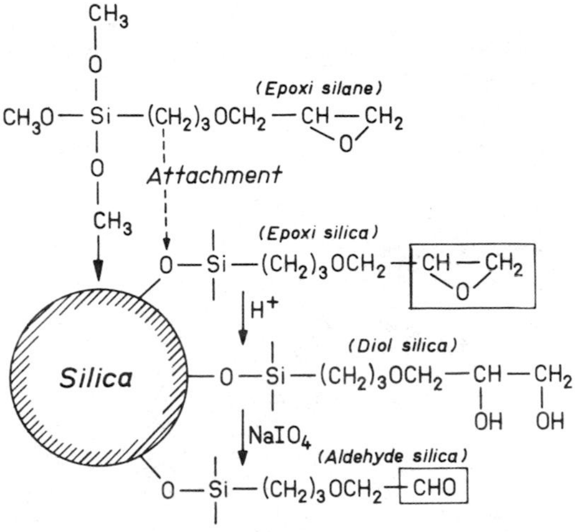

FIGURE 19.1 Activation of silica for HPLAC.

TABLE 19.1 Application of High-Performance Liquid Affinity Chromatography

Ligand	Active group for ligand coupling	Application
Cibacron Blue F3G-A	Epoxy	Dehydrogenases, kinases, ribonuclease
NAD^+		Lactate dehydrogenase
Adenosine monophosphate	Aldehyde	Dehydrogenases including isoenzymes
Anti-human serum albumin		Albumin
Anti-human creatine kinase		Isoenzymes of creatine kinase
Concanavalin A		Glycoproteins

Source: From Mosbach et al. (1982).

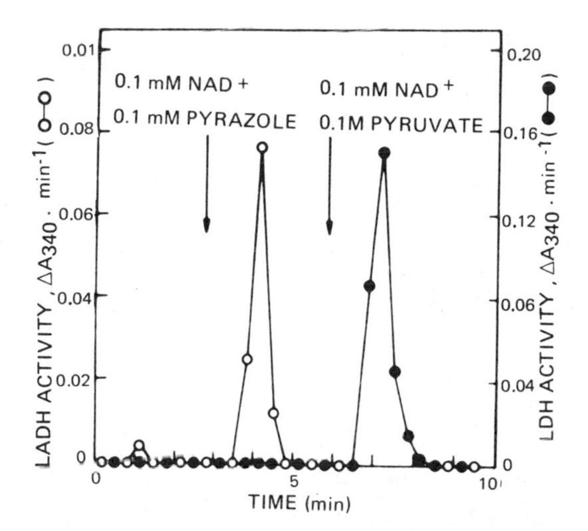

FIGURE 19.2 Separation of liver alcohol dehydrogenase (LADH) and lactate dehydrogenase (LDH) from pig heart with ternary complex formation on an AMP-silica column. A mixture (10 μl) of LDH (10 μg) and LADH (20 μg) dissolved in 0.1 M sodium phosphate buffer, pH 7.5, was injected onto the column and eluted, as indicated by the arrows, with pulses of 1.6 ml. Flow rate was 1.5 ml/min, and pressure, 400 psi. (From Larsson et al., 1979.)

AMP-silica gel and biospecifically eluted and separated under the conditions given in Figure 19.2 within 10 min.

HPLAC separation was successfully performed using other adsorbent materials with sufficient mechanical stability, such as Separon 1000 (Čoupek, 1982). When it is modified by ω-aminocaproyl-L-Phe-D-Phe-OCH_3, good analytic characterization of trypsin preparations within 15 min (concerning the desorption step) is possible.

Hybrid silica gels with a low ratio of hydrophobic to hydrophilic groups have enabled the separation of several test proteins, such as insulin, egg white lysozyme, myoglobin, and β-lactoglobulin in the sense of a high-pressure hydrophobic chromatography (Nishikawa et al., 1982).

The present literature reported in the field of HPLAC show clearly that the application of this method for separation of biomacromolecules on analytic and semipreparative scale is still in a pioneering stage. Further progress may be achieved in the future by optimization of the adsorbent materials and the required technical conditions.

Finally, enlargement of HPLAC can be expected by introduction of further affinity methods, such as metal chelate (Chap. 13) and charge transfer adsorption chromatography (Chap. 19).

REFERENCES

Čoupek, J. (1982). Macroporous spherical hydroxyethylmethacrylate copolymers, their properties, activation and use in high performance affinity chromatography. In *Affinity Chromatography and Related Techniques*, T. C. J. Gribnau, J. Visser, and R. J. F. Nivard (eds.). Elsevier Scientific, Amsterdam, pp. 165-179.

Larsson, P.-O., Griffin, T., and Mosbach, K. (1979). Some new techniques related to affinity chromatography. In *Affinity Chromatography and Molecular Interactions*, J.-M. Egly (ed.). Editions INSERM, Paris, pp. 91-97.

Mosbach, K., Glad, M., Larsson, P.-O., and Ohlson, S. (1982). Affinity precipitation and high performance liquid affinity chromatography. In *Affinity Chromatography and Related Techniques*, T. C. J. Gribnau, J. Visser, and R. J. F. Nivard (eds.). Elsevier Scientific, Amsterdam, pp. 201-205.

Nishikawa, A. H., Roy, S. K., and Puchalski, R. (1982). High-pressure hydrophobic chromatography of proteins. In *Affinity Chromatography and Related Techniques*, T. C. J. Gribnau, J. Visser, and R. J. F. Nivard (eds.). Elsevier Scientific, Amsterdam, pp. 471-482.

Ohlson, S., Hansson, L., Larsson, P.-O., and Mosbach, K. (1978). High performance liquid affinity chromatography (HPLAC) and its application to the separation of enzymes and antigens. *FEBS Lett.* *93*: 5-9.

Regnier, F. E., and Noel, R. (1976). Glyceropropylsilane bonded phases in the steric exclusion chromatography of biological macromolecules. *J. Chromatogr. Sci.* *14*: 316-320.

APPENDIX

Concluding Remarks

The aim of the preceding chapters was to give a general survey of the most essential areas of affinity chromatography, including several theoretical aspects, and to describe the present state of application of this method as a tool for separation, purification, and analytic detection of biologically active molecules and substances. Since at present and in the near future total synthesis is out of the question for most biologically active substances – and this is also valid for biomacromolecules – they must be isolated from biological materials, such as microorganisms, plants, or selected organs of animals. The desired substances, however, normally occur in nature in low concentrations and combined with many other materials. Separation and sufficient purification are then possible only by expensive multistep procedures with a more or less great loss of substance and biological activity. Therefore, the scientists in this field are constantly striving to improve the separation procedures and to develop new and more efficient methods. As such, affinity chromatography has proved a favorable and versatile variant, and the rapid development in this field demonstrates the progress attained. A representative example underlining this statement is given by the isolation of the insulin receptor of liver cells (Cuatrecasas, 1972). The traditional multistep procedure permits its 60-fold enrichment, whereas affinity chromatography using insulin-agarose as adsorbent allows a 800-fold purification in one step.

In all probability it can be assumed that the development of the general principles of affinity chromatography concerning its application on laboratory scale has now reached a final stage. The great attraction the method has gained in the last few years in its application on a small scale now raises the question whether and to what extent it is also suitable for the preparation of biologically active compounds in larger quantities, such as on a pilot or industrial scale. The pioneering work along these lines has shown that such procedures require the clarification of several new and special problems that have been investigated so far only in part. Since in scale-up preparations columns of essentially higher dimensions than usual on a laboratory scale are used, the main problem concerns the adsorbent material, which must be of excellent mechanical stability to guarantee good flow rates in high-volume columns applied in scale-up procedures. Because the scale-up methods developed up to now concern mainly procedures for preparations of clinically interesting biosubstances, such as shown in Table 1, moreover, a sterilization must be practicable, and for this reason heat stability of the adsorbent material is required. Further demands are a certain macroporosity, the presence of groups suitable for the introduction of required ligands, minimal nonspecific adsorptions, and low cost of the adsorbent material.

When the substance to be isolated by affinity chromatography is to be prepared for clinical use, prevention of adsorbent leakage leading to impurities of the eluate and consequently of the product is one of the most essential problems to be solved. Eketorp (1982) has found in using radiolabeled glycine coupled to Sepharose CL-6B that the total leakage of the ligand was approximately 10% of coupled ligand, and it was greatly increased when using buffers containing nucleophilic groups or molecules, such as NH_3. The example demonstrates impressively that leakage processes can disturb the separation when common gels are used and that the applicability of a method is called into question if this effect cannot be prevented. Unfortunately, there exists no general concept up to now to solve this problem, but each procedure must be treated more or less individually. In the case of poliovirus isolation by Sepharose 4B columns, for example, the leakage of antibodies could be widely restricted by using Sepharose CL-4B and additionally cross-linking of the antibodies with glutaraldehyde (van der Wezel and van der Marel, 1982; Table 1). Principally it should be stated, however, that the leakage problem is not yet solved satisfactorily. A further demand is the sterility of the columns when biosubstances for clinical use are to be produced. This requires autoclavable gels, that is, those that are widely heat stable. This problem can also be overcome by chemical sterilization. van der Wezel and van der Marel (1982) attained sterility of their antibody-Sepharose columns by storage in a 0.5% phenol solution that could be easily removed by thorough washing before use. It must be emphasized, however, that much work will have to be spent in this field of application.

TABLE 1 Examples for Large-Scale Affinity Chromatography of Clinical Interesting Biosubstances

Substance	Adsorbent	Sample volume	Elution	References
Albumin	Blue Trisacryl[a]	50 liters	50 mM tris·HCl + 3.5 M NaCl, pH 8	Saint-Blancard et al. (1982)
Bovine and human fibrinoectins	Gelatin agarose[b]	Limited to 8 liters plasma	4 M urea, 50 mM tris buffer, pH 7.5	Roulleau et al., (1982)
Poliovirus type 1	Antibodies coupled to Sepharose 4B[c]	—	5-7 M NH_4SCN	Van der Wezel and van der Marel (1982)

[a]Cibacron Blue F3G-A coupled to highly cross-linked copolymer of N-tris(hydroxymethyl)methylacrylamide.
[b]Prepared by mixing and polymerization of agarose with gelatin followed by cross-linking with glutaraldehyde.
[c]A leakage of antibodies can be overcome by using Sepharose CL-4B and additional cross-linking of the antibodies with glutaraldehyde.

Besides the scale-up preparation of biologically active substances of clinical interest, one can assume that procedures will be increasingly developed in the future for the preparation of substances for use in other fields of research and practice. This concerns, among other things, (a) the specific isolation of antibodies from culture media, (b) biochemicals for genetic engineering, (c) radioisotopically labeled affinity-purified antibodies for location and therapy of certain tumors, and (d) conjugates for the immunoassay technique (Chap. 8) (Eveleigh, 1982).

By reason of the low price of the ligands and the simplicity of their coupling to the adsorbent particles, one can assume that dye-ligand chromatography will also gain increasing significance in the future as a scale-up method.

Another problem is the theoretical description of processes closely connected with affinity chromatography. This concerns in particular the behavior of biomacromolecules, viruses, and cells on the interface of the solid phase (the adsorbent particles) with the liquid phase (the sample solution as well as the washing and elution buffer) under conditions of chromatography. Problems resulting from this matter concern correlations between surface tension of the adsorbent material and the stability of the adsorbed substance, the nature and the distribution of chemical groups on the adsorbent surface and unspecific adsorption of substances contained as contaminations in the sample solution, kinetic problems of the adsorption and elution step, and others. There is no doubt that the clarification of such questions can provide essential conditions to optimize the efficiency of materials. Such investigations are just beginning.

We have emphasized the importance of affinity chromatography as an essential premise for the deepening of our knowledge of living processes on a molecular level in the last few years. On the other hand, the development of new procedures and principles in affinity techniques, including analytic methods, has been induced, as was shown in some of the preceding chapters.

It can be expected, furthermore, that the prospective development of more and more large-scale procedures in affinity chromatography will lead to increased industrial and pilot production of many new biologically active substances, which will then serve as an essential source for further progress in biochemical research work.

REFERENCES

Cuatrecasas, P. (1972). Affinity chromatography and purification of the insulin receptor of liver cell membranes. *Proc. Nat. Acad. Sci. U. S. 69*: 1277-1281.

Eketorp, R. (1982). Affinity chromatography in industrial blood plasma fractionation. In *Affinity Chromatography and Related Techniques*, T. C. J. Gribnau, J. Visser, and R. J. F. Nivard (eds.). Elsevier Scientific, Amsterdam, pp. 263-273.

Eveleigh, J. W. (1982). Practical considerations in the use of immunosorbents and associated instrumentation. In *Affinity Chromatography and Related Techniques*, T. C. J. Gribnau, J. Visser, and R. J. F. Nivard (eds.). Elsevier Scientific, Amsterdam, pp. 293-303.

Saint-Blancard, J., Kirzin, J. M., Riberon, P., Petit, F., Fourcart, J., Girot, P., and Boschetti, E. (1982). A simple and rapid procedure for large scale preparation of IgG's and albumin from human plasma by ion exchange and affinity chromatography (1982). In *Affinity Chromatography and Related Techniques*, T. C. J. Gribnau, J. Visser, and R. J. F. Nivard (eds.). Elsevier Scientific, Amsterdam, pp. 305-312.

Roulleau, M. F., Boschetti, E., Bournouf, J. M. Kirzin, J. M., and Saint-Blancard, J. (1982). Bovine and human fibrinoectins: Large-scale preparation by affinity chromatography. In *Affinity Chromatography and Related Techniques*, T. C. J. Gribnau, J. Visser, and R. J. F. Nivard (eds.). Elsevier Scientific, Amsterdam, pp. 323-331.

Van der Wezel, A. L., and van der Marel, T. P. (1982). The application of immuno-adsorption on immobilized antibodies for large-scale concentration and purification of vaccines. In *Affinity Chromatography and Related Techniques*, T. C. J. Gribnau, J. Visser, and R. J. F. Nivard (eds.). Elsevier Scientific, Amsterdam, pp. 283-292.

Index

B

C

R

S

T